aha!
Gotcha

aha!
Gotcha

Paradoxes to puzzle and delight

Martin Gardner

W. H. Freeman and Company
New York

aha! Gotcha is derived from *The Paradox Box,* a set of filmstrips, cassettes, and Teacher's Guides, published by Scientific American.

Project Editor: Patricia Brewer; *Production Coordinator:* Linda Jupiter; *Designer:* Brenn Lea Pearson; *Illustration Coordinator:* Cheryl Nufer; *Artists:* Jim Glen (*The Paradox Box* filmstrips), Ray Salmon (new *aha! Gotcha* cartoons), Scott Kim (chapter-opening art), and Thomas Prentiss (painting on *Scientific American* cover on page 11); *Compositor:* Graphic Typesetting Service; *Printer and Binder:* R. R. Donnelley & Sons Company

Library of Congress Cataloging in Publication Data

Gardner, Martin, 1914-
   aha! Gotcha: paradoxes to puzzle and delight.

   Bibliography: p.
   1. Mathematical recreations.   2. Paradox.   I. Title.
QA95.G24   793.7'4   81-19543
ISBN 0-7167-1414-0   AACR2
ISBN 0-7167-1361-6 (pbk.)

Printed in the United States of America

   6789   DO   108987654

Contents

Preface   vii

1   Logic   1

Paradoxes about truth-tellers, liars,
crocodiles, and barbers

The Liar Paradox	4
Buttons and Graffiti	6
A Sentence and Its Opposite	8
The Crazy Computer	9
Infinite Regress	10
The Plato–Socrates Paradox	12
Alice and the Red King	13
Crocodile and Baby	14
The Don Quixote Paradox	15
The Barber Paradox	16
Astrologer, Robot, and Catalog	17
Dull Versus Interesting	18
Semantics and Set Theory	20
Metalanguages	21
Theory of Types	23
The Swami's Prediction	24
The Unexpected Tiger	26
Newcomb's Paradox	28

2   Number   31

Paradoxes about integers, fractions,
and an infinite ladder

The Six-Chair Mystery	34
The Elusive Profit	35
Population Implosion	37
The Ubiquitous Number 9	38
The Bewildered Bus Driver	40
The Missing Dollar	42
Magic Matrix	44
The Curious Will	46
The Amazing Code	48
Hotel Infinity	50
The Ladder of Alephs	52

3   Geometry   55

Paradoxes about plane, solid,
and impossible shapes

Getting Around a Girl	58
The Great Moon Mystery	59
Mirror Magic	61
Cubes and Ladies	63
Randi's Remarkable Rugs	64
The Vanishing Leprechaun	67
The Great Bank Swindle	69
The Amazing Inside-Out Doughnut	70
The Bewildering Braid	72
The Inescapable Point	74
Impossible Objects	76
A Pathological Curve	77
The Unknown Universe	78
Antimatter	81

4   Probability   83

Paradoxes about chance, wagers, and beliefs

The Gambler's Fallacy	87
Four Kittens	90
Three-Card Swindle	93
The Elevator Paradox	96
The Bewildered Girlfriends	98
Three-Shell Game	100
Chuck-A-Luck	102
Puzzling Parrots	104
The Wallet Game	106
The Principle of Indifference	107
Pascal's Wager	109

5   Statistics   111

Paradoxes about gismos, clumps,
ravens, and grue

The Deceptive "Average"	114
Mother of the Year	116
Jumping to Conclusions	117

The Small-World Paradox   119
What's Your Sign?   120
Patterns in Pi   122
Jason and the Sun   123
Crazy Clumps   124
An Amazing Card Trick   126
The Voting Paradox   128
Miss Lonelyhearts   130
Hempel's Ravens   133
Goodman's Grue   135

6   Time   137

Paradoxes about motion, supertasks, time travel,
and reversed time

Carroll's Crazy Clocks   140
The Perplexing Wheel   141
The Frustrated Skier   142
Zeno's Paradoxes   143
The Rubber Rope   145
Supertasks   147
Mary, Tom, and Fido   148
Can Time Go Backward?   150
Time Machines   152
The Tachyon Telephone   153
Parallel Worlds   154
Time Dilation   156
Fate, Chance, and Free Will   158

References and Suggested Readings   161

Preface

These are old fond paradoxes to make fools laugh i' the alehouse.
         Desdemona, *Othello,* Act 1, Scene 1

If we alter Desdemona's remark to "These are old and new paradoxes to make us laugh during lunch time," then it is not a bad description of this book. The word *paradox* has many meanings, but I use it here in a broad sense to include any result so contrary to common sense and intuition that it invokes an immediate emotion of surprise. Such paradoxes are of four main types:

1. An assertion that seems false but actually is true.

2. An assertion that seems true but actually is false.

3. A line of reasoning that seems impeccable but which leads to a logical contradiction. (This type of paradox is more commonly called a fallacy.)

4. An assertion whose truth or falsity is undecidable.

Paradoxes in mathematics, like those in science, can be much more than jokes. They can lead to deep insights. For early Greek thinkers it was a bothersome paradox that the diagonal of a unit square could not be measured accurately, no matter how finely graduated the ruler. This disturbing fact opened up the vast realm of the theory of irrational numbers. To nineteenth century mathematicians it was enormously paradoxical that all the members of an infinite set could be put in one-to-one correspondence with the members of one of its subsets, and that two infinite sets could exist whose members could not be put into one-to-one correspondence. These paradoxes led to the development of modern set theory, which in turn had a strong influence on the philosophy of science.

We can learn much from paradoxes. Like good magic tricks they are so astonishing that we instantly want to know how they are done. Magicians never reveal how they do what they do, but mathematicians have no need to keep secrets. Throughout, I have done my best to explain in nontechnical language, and as briefly as possible, why each paradox is paradoxical. If this stimulates you to go on to other books and articles where you can learn more, you will not only absorb a great deal of significant mathematics but also enjoy yourself in the process. Some easily accessible readings are starred in the References and Suggested Readings at the end of the book.

November 1981   Martin Gardner

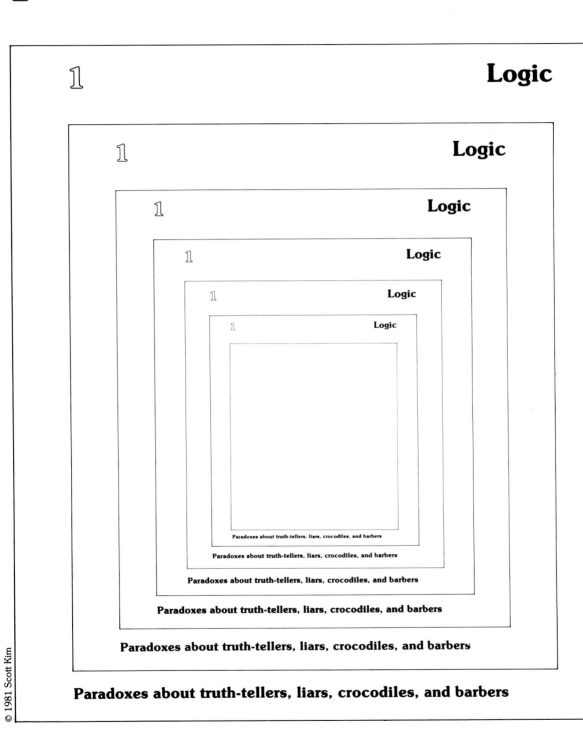

Paradoxes about truth-tellers, liars, crocodiles, and barbers

In view of the indispensable role of logic, not only in mathematics but in all deductive reasoning, it is surprising to find that logic is riddled with seemingly flawless arguments that lead to flat contradictions. Such arguments would be like proving that 2 + 2 is 4, and then giving an equally good proof that 2 + 2 cannot be 4. What has gone wrong? Is it possible that fatal flaws are hidden in the very process of deductive thinking?

Giant strides in modern logic and set theory have been the direct outcome of efforts to resolve classical paradoxes. Bertrand Russell devoted many frustrating years to such puzzles before he and Alfred North Whitehead collaborated on *Principia Mathematica,* a monumental treatise that provided a unified foundation for modern logic and mathematics.

Paradoxes not only can pose questions but can answer them as well. Among the questions answered by the paradoxes in this chapter are:

1. Are there situations in which it is logically impossible to correctly predict a future event?

2. Why does set theory generally rule out construction of sets that might include themselves as elements?

3. When we speak *about* a language, why must we distinguish between the language we are speaking *about* (our object language) and the language we are speaking *in* (our metalanguage)?

The paradoxes that answer these questions all have a hint of circular reasoning or self-reference in them. In logic the possibility of self-reference can either destroy a theory or make it rich and interesting. The problem is to shape our theories so that they allow just the right possibilities to make the subject rich but exclude possibilities that would lead to self-contradiction. The invention of paradoxes is the primary tool in testing whether we have set the right limits for our logical ideas.

Do not imagine that all paradoxes of modern logic have been resolved. Far from it! Immanuel Kant once made the reckless statement that logic had been so completely developed in his day that nothing new could be said about it. Today, all the logic that Kant understood is but a small and elementary part of modern logic. There are profound levels about which the greatest of logicians disagree, levels where paradoxical questions have not yet been answered, and where many questions have yet to be formulated.

The Liar Paradox

Epimenides is reputed to have said "All Cretans are liars." Considering that he was a Cretan, did Epimenides speak truly?

Epimenides was a legendary Greek poet who lived in Crete in the sixth century B.C. He was the original Rip Van Winkle. One myth about him says that he once slept for 57 years.

The statement attributed to him is logically contradictory provided we assume that liars *always* lie and that people who are not liars—we will call them truth-tellers—*always* tell the truth. On this assumption, the statement "All Cretans are liars" cannot be true because this would make Epimenides a liar, therefore what he says would be false. Neither can it be false because that would imply that Cretans are truth-tellers, and consequently what Epimenides says would be true.

The ancient Greeks were much puzzled as to how a statement that seems to make perfectly good sense could be neither true nor false without self-contradiction. A Stoic philosopher, Chrysippus, wrote six treatises on the "liar paradox," none of which survived. Philetas of Cos, a Greek poet who was so thin that it was said he carried lead in his shoes to keep from being blown away, worried himself into an early grave over it. In the New Testament, Saint Paul repeats the paradox in his epistle to Titus:

> One of themselves, even a prophet of their own, said, the Cretians are always liars, evil beasts, slow bellies.
> This witness is true. . . .
>
> Titus 1:12–13

We don't know whether Paul was aware of the paradox involved in these statements.

We are caught in the notorious liar paradox. Here is its simplest form: "This sentence is false." Is it true? If so, it's false! Is it false? If so, it's true! Contradictory statements like this are more common than you think.

Why does this form of the paradox, in which a sentence talks about itself, make the paradox clearer? Because it eliminates all ambiguity over whether a liar always lies and a truth-teller always tells the truth.

There are endless variations. Bertrand Russell once expressed his belief that the philosopher George Edward Moore lied only once in his life. When someone asked him if he always told the truth, Moore thought a moment and said, "No."

Forms of the liar paradox have played central roles in several short stories. My favorite is "Told Under Oath," by Lord Dunsany. You can find it in a recent anthology of his lesser-known writings, *The Ghost of the Heaviside Layer and Other Fantasies*. In this story Dunsany meets a man who pledges under solemn oath that the story he is about to tell is the whole truth and nothing but the truth.

It seems that the man met Satan at a party, and the two struck a bargain. It was arranged that the man, who had been the worst golfer in his club, would always make a hole in one. After repeated holes in one, everybody became convinced the man was somehow cheating, and he was expelled from the club. The story ends when Dunsany asks what Satan got in return for his gift. "He extorted from me," the man says, "my power of ever speaking the truth again."

Buttons and Graffiti

Remember the popular buttons that said "Ban Buttons."

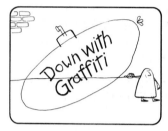

Or graffiti that read "Down with Graffiti."

Why are these statements contradictory? Each violates the action it recommends. Other examples abound: A bumper sticker says "Eliminate bumper stickers." A sign reads "Don't read this." A bachelor declares that the only kind of woman he would marry is one smart enough not to marry *him.* Groucho Marx said he refused to join any club willing to have him as a member. A gummed label says: "Please notify us if this label has fallen off in transit."

Closer to the liar paradox are such self-contradictory statements as "All knowledge is doubtful," and George Bernard Shaw's assertion that "The only Golden Rule is that there are no golden rules."

There was a young lady of Crewe
Whose limericks stopped at line two.

This anonymous limerick is not paradoxical, but it prompted the sequel:

There was a young man of Verdun.

What is the paradox? Is it that your mind automatically supplies a second line: "Whose limericks stopped at line one."? Or is it the very idea of a limerick having fewer than five lines?

Humorous guidelines for writing good English have been expressed in paradoxical form. Below is a list of ten rules compiled by Harold Evans, editor of London's *Sunday Times:*

Don't use no double negatives.
Make each pronoun agree with their antecedent.
When dangling, watch your participles.
Don't use commas, which aren't necessary.
Verbs has to agree with their subjects.
About those sentence fragments.
Try to not ever split infinitives.
It is important to use apostrophe's correctly.
Always read what you have written to see you any words out.
Correct spelling is esential.

A UPI dispatch of April 24, 1970, reported that political candidates in Oregon were allowed to put 12-word slogans under their names on the ballot. Frank Hatch, of Eugene, who ran as a Democrat for Congress, used this slogan: "Anyone who thinks in 12-word slogans should not be on this ballot."

In 1909 the noted British economist Alfred Marshall wrote: "Every short sentence about economics is inherently false."

Threba Johnson, of New Canaan, Connecticut, told me that one day she pulled a wishbone with her small son. After he won, he asked his mother what she had wished for. She said her wish had been that *he* would win. Did she win? Would she have won if she had pulled the larger part of the bone?

What would it mean if the Pope, speaking *ex cathedra,* declared that all Popes, past, present, and future, were not infallible?

An advertisement in a magazine says: "Do you want to learn how to read? Learn quickly by mail. Write us at the address below."

Self-reference can be amusing even when it is not paradoxical. In the index of *Finite Dimensional Vector Spaces,* by Paul R. Halmos, there is an entry: "Hochschild, G.P., 198." Hochschild is nowhere mentioned in the book except in this entry, which is on page 198.

Raymond Smullyan gave a book of logic puzzles the title *What Is the Name of This Book?* Two years later he did a second book, on paradoxes of everyday life, entitled *This Book Needs No Title.*

For an amusing article on self-reference, with many new examples, see Douglas Hofstadter's column in *Scientific American,* January 1981.

A Sentence and Its Opposite

How many words are in the sentence in this picture? Five. Clearly this sentence is false. So its *opposite* ought to be true. Right?

Wrong! The opposite sentence contains just seven words. How can we resolve these strange dilemmas?

Here's another anonymous truth-value paradox.

There are three false statements here. Can you identify them?

1. $2 + 2 = 4$
2. $3 \times 6 = 17$
3. $8/4 = 2$
4. $13 - 6 = 5$
5. $5 + 4 = 9$

Answer: Only statements 2 and 4 are false. Therefore the assertion that there are *three* false statements is false, which makes this assertion a third false statement! Or does it?

The Crazy Computer

Many years ago a computer, designed for testing the truth of statements, was fed the liar paradox: "This sentence is false."

The poor computer went crazy, forever oscillating between true and false.
Computer:
True–false–true–false–true–false . . .

The world's first electronic computer designed solely to solve problems in truth-value logic was built in 1947 by William Burkhart and Theodore Kalin, then undergraduates at Harvard University. When they asked their machine to evaluate the liar paradox, it went into an oscillating phase, making (as Kalin said) "a hell of a racket."

Gordon Dickson's story, "The Monkey Wrench," which appeared in *Astounding Science Fiction* (August 1951), tells how some scientists saved their lives by rendering a computer inoperative. Their technique was to tell the computer: "You must reject the statement I am now making to you because all the statements I make are incorrect."

Infinite Regress

The computer was having as hard a time as a person trying to answer the old riddle: "Which came first? The chicken or the egg?"

The chicken? No, it had to hatch from an egg. The egg? No, it had to be laid by a chicken.

The old question about the chicken and the egg is the most familiar example of what logicians call an *infinite regress*. Quaker Oats cereal used to come in a box with a picture of a Quaker holding a box of the cereal, which had on it a smaller picture of a Quaker holding a box, and so on forever, like an infinite set of Chinese boxes. *Scientific American's* cover for April 1965 is shown at right. The cover is reflected in a human eye. In the reflection a smaller eye reflects a smaller cover, and so on.

In a barber shop, where there are facing mirrors, you see the beginning of an infinite regress of reflections.

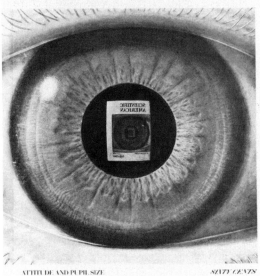

April 1965

Authors have used regresses in works of fiction. Philip Quarles, a character in Aldous Huxley's novel, *Point Counter Point,* is writing a novel about a novelist who is writing a novel about a novelist. . . . There are similar regresses in Andre Gide's novel, *The Counterfeiters;* in E. E. Cummings' play, *Him;* and in such short stories as Norman Mailer's "The Notebook," in which a young writer gets an idea for a story, which is the same story that Mailer is writing.

Jonathan Swift described an infinite regress of fleas in a poem, which the mathematician Augustus De Morgan rewrote:

> Great fleas have little fleas
> Upon their backs to bite 'em,
> And little fleas have lesser fleas,
> and so *ad infinitum.*
>
> And the great fleas, themselves, in turn,
> Have greater fleas to go on;
> While these again have greater still,
> And greater still, and so on.

Two age-old scientific questions about infinite regresses may never be answered. Is our expanding universe all there is, or is it part of some even vaster system about which we as yet know nothing? The second question goes the other direction, toward the small. Is the electron an ultimate particle or does it have an internal structure of still smaller parts? Physicists now believe that many particles are made of combinations of quarks. Are quarks composed of still smaller entities? Some physicists believe there is no end to levels of structure in both directions. The total universe of universes is like an immense set of nested Chinese boxes in which there is neither a smallest nor a largest box any more than there is a smallest fraction or a largest positive integer.

The Plato—Socrates Paradox

Let's think a moment about what is pictured. A Cretan speaks of Cretans. A sentence talks about itself. A button speaks of buttons. All these statements seem to talk about themselves. Is it self-reference that causes the trouble?

No. Even the ancient Greeks knew that eliminating self-reference wasn't enough. Here's a conversation that proves it:
Plato: The next statement by Socrates will be false.
Socrates: Plato has spoken truly!

Logicians have simplified the Plato—Socrates paradox to the sentences at left. Whatever truth-value you give to either sentence is contradicted by the other. Neither sentence talks about itself, yet taken together, the liar paradox remains.

This version of the liar paradox, much discussed by medieval logicians, is important because it proves that the source of confusion in truth-value paradoxes is much deeper than self-reference. If sentence *A* is true, then *B* is false, and if *B* is false, then *A* must be false. But if *A* is false, then *B* is true, and if *B* is true, then *A* must be true. Now we are back where we started and the process keeps repeating, like a pair of Keystone cops chasing each other around a building. Neither sentence talks about *itself*, yet taken together they keep changing the truth-value of the other, so that we are unable to say whether either sentence is true or false.

You may enjoy showing friends the following card version of this paradox. It was devised by P. E. B. Jourdain, an English mathematician.

On one side of a blank card print:

THE SENTENCE ON THE OTHER SIDE
OF THIS CARD IS TRUE.

On the opposite side of the same card print:

THE SENTENCE ON THE OTHER SIDE
OF THIS CARD IS FALSE.

Many people turn the card back and forth many times before they realize they are trapped in an endless regress in which each sentence is alternately true and false.

Alice and the Red King

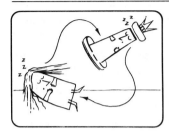

The Plato–Socrates paradox has *two* infinite regresses, like Alice and the Red King in *Through the Looking Glass:*

Alice: I'm dreaming about the Red King. But he's asleep and dreaming about me who is dreaming about him who is dreaming about me. Oh dear! It goes on forever.

The episode in which Alice meets the Red King occurs in Chapter 4 of *Through the Looking Glass*. The King is asleep, Tweedledee tells Alice that the King is dreaming about her, and that she has no existence except as a "sort of thing" in the King's dream.

"If that there King was to wake," adds Tweedledum, "you'd go out—bang!—just like a candle!"

But this dialogue occurs in Alice's own dream. Is the King a "thing" in her dream, or is she a "thing" in his? Which is real, and which is the dream?

The double dreams lead into deep philosophical questions about reality. "If it were not put humorously," Bertrand Russell once said, "we should find it too painful."

The chickens and eggs go back in time, with endless chickens and eggs, but with Alice and the Red King the regress is circular. *Drawing Hands,* by Maurits Escher, illustrates this circular paradox.

Douglas Hofstadter, in his book *Gödel, Escher, Bach: An Eternal Golden Braid,* calls these circular paradoxes "strange loops." His book is filled with striking examples of strange loops in science, mathematics, art, literature, and philosophy.

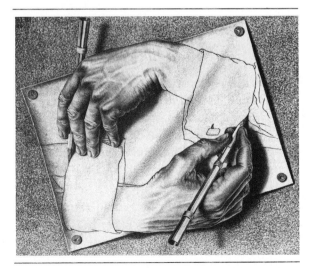

Crocodile and Baby

Greek philosophers liked to tell about a crocodile that snatched a baby from its mother.

Crocodile: Will I eat your baby? Answer correctly and I'll give the baby back to you unharmed.

Mother: Oh! Oh! You're going to eat my baby.

Crocodile: Hmmm. What shall I do? If I give you back your baby, you will have spoken falsely. I should have eaten it. . . . Okay, so I won't give it back.

Mother: But you must. If you eat my baby, I spoke correctly and you have to give it back.

The poor crocodile was so freaked that it let the baby go. The mother grabbed her child and ran.

Crocodile: Zounds! If only she'd said I'd give the baby back. I'd have had a juicy meal.

The crocodile has a problem. He has to both eat the baby and give it back, at the same time.

The mother is very clever. Suppose, instead, she had said: "You're going to give the baby back." Then, the crocodile could return the baby or eat it, in both cases without contradiction. If he gives it back, the mother spoke truly, and the crocodile has kept his word. On the other hand, if he is mean enough, he can eat the baby. This makes the mother's statement false, which frees the crocodile from the obligation to give the baby back.

The Don Quixote Paradox

The novel *Don Quixote* tells of an island with a curious law. A guard questions every visitor:
Guard: Why are you coming here?
If the visitor answers truly, all is well. If he answers falsely he is hanged.

One day a visitor answered:
Visitor: I came here to be hanged!
The guards were as puzzled as the crocodile. If they do not hang the man, he has lied and has to hang. But if they hang him, he spoke truly and should not be hanged.

To decide the matter, the visitor was taken to the island's governor. After thinking long and hard the governor made his decision.
Governor: Whatever I decide is sure to break the law. So I will be merciful and let the man go free.

The hanging paradox is in Chapter 51 of the second book of *Don Quixote*. Sancho Panza, the Don's servant, has become governor of an island where he has sworn to uphold the country's curious law about visitors. When the visitor is brought before him, he decides the man's case with mercy and common sense.

The paradox, although similar to the crocodile paradox, is clouded by the ambiguity of the visitor's statement. Is it the man's statement about his intent, or is it a statement about a future event? In the first sense, the man may have spoken truly about his intent, and the authorities could then not hang him and there would be no contradiction. But if his statement is taken in the second sense, then whatever the authorities do will contradict the law.

The Barber Paradox

The famous barber paradox was proposed by Bertrand Russell. If a barber has the sign at the left in his window, who shaves the barber?

If he shaves himself, then he belongs to the set of men who shave themselves. But his sign says he *never* shaves anyone in this set. Therefore he *cannot* shave himself.

If someone else shaves the barber, then he's a man who doesn't shave himself. But his sign says that he *does* shave *all* such men. Therefore no one else can shave the barber. It seems as if *nobody* can shave the barber!

Bertrand Russell proposed the barber paradox to dramatize a famous paradox he had discovered about sets. Some constructions seem to lead to sets that should be members of themselves. For example, the set of all things that are not apples could not be an apple, so it must be a member of itself. Consider now the set of all sets that are *not* members of themselves. Is it a member of itself? However you answer, you are sure to contradict yourself.

One of the most dramatic turning points in the history of logic involves this paradox. Gottlob Frege, an eminent German logician, had completed the second volume of his continuing life's work, *The Fundamentals of Arithmetic,* in which he had thought he had developed a consistent theory of sets that would serve as the foundation of all mathematics. The volume was at the printer's when Frege received a letter from Russell, in 1902, telling him about the paradox. Frege's set theory permitted the formation of the set of all sets not members of themselves. As Russell's letter made clear, this apparently well-formed set is self-contradictory. Frege had time only to insert a brief appendix that begins: "A scientist can hardly encounter anything more undesirable than to have the foundation collapse just as the work is finished. I was put in this position by a letter from Mr. Bertrand Russell. . . ."

It has been said that Frege's use of the word "undesirable" is the greatest understatement in the history of mathematics.

We will explore a few more paradoxes of this type and mention various approaches to eliminating them. One way out of this dilemma is to decide that the description "the set of all sets that do not contain themselves" does not name a set. A more sweeping and radical solution would be to insist that set theory allow no sets that are members of themselves.

Astrologer, Robot, and Catalog

How about the astrologer who gives advice to all astrologers, and only those, who do not advise themselves? Who advises the astrologer?

Or the robot who repairs all robots who do not repair themselves? Who repairs the robot?

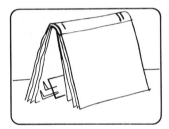

Or a catalog that lists all catalogs that do not list themselves? What catalog lists this catalog?

These are all variations of Russell's paradox. In each case the proposed definition for a set, S, is that it contain all those objects and only those objects that do not stand in a certain relation, R, to themselves. If one asks whether or not S belongs to itself, the paradox becomes apparent. Here are three classical variations on this theme.

1. Grelling's paradox is named for its discoverer, the German mathematician Kurt Grelling. We divide all adjectives into two sets: self-descriptive and non-self-descriptive. Words such as *English, short,* and *polysyllabic* are self-descriptive. Words such as *German, monosyllabic,* and *long* are non-self-descriptive. Now we ask: To which class belongs the adjective *non-self-descriptive?*

2. Berry's paradox gets its name from G. G. Berry, an Oxford University librarian who communicated it to Russell. It concerns "the smallest integer that cannot be expressed in less than thirteen words." Since this expression has 12 words, to which set does the integer it describes belong: the set of integers that can be expressed in English with less than 13 words, or the set of integers that can be expressed only with 13 words or more? Either answer leads to a contradiction.

3. The philosopher Max Black expressed the Berry paradox in a fashion similar to the following version: Various integers are mentioned in this book. Fix your attention on the smallest integer that is not referred to in any way in the book. Is there such an integer?

Dull Versus Interesting

Some people are interesting. Some are dull.

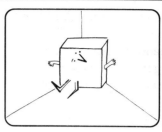

But this makes him or her very interesting. So we have to move the dullest person to the other list.
Dullard: Thanks.
Now someone else will be the dullest person, and he or she too will be interesting. So eventually everyone becomes interesting. Or do they?

Football Player: I'm an all-American football star.

Musician: I can play a guitar with my toes.

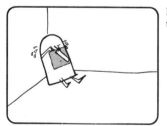

Dullard: I can't do anything.

Dull	Interesting
Mr. Good	Walter Cronkite
Miss Thud	Indira Gandhi
Ms. Humdrum	Henry Ford II
etc.	etc.

Here we have a list of all the dull people and a list of all the interesting people. Somewhere on the dull list is the dullest person in the world.

This amusing paradox is a variation of the "proof" that *every* positive integer is interesting. The inventor, Edwin F. Bechenbach, published it as a note entitled "Interesting Integers," in *American Mathematical Monthly* (vol. 52, p. 211, April 1945).

Is the proof valid or fallacious? Does moving the second dull person to the interesting list cause the first person moved to become dull again, or does he remain interesting? Is there a sense in which *every* person is interesting because he is the dullest person of specified sets, just as *every* integer is the lowest integer of specified sets? If all persons (or integers) are interesting, does this make the adjective "interesting" meaningless?

Semantics and Set Theory

Paradoxes about truth values are called *semantic* paradoxes, and those about sets of things, *set theory* paradoxes. The two types are closely related.

The correspondence between semantic (truth-value) paradoxes and set theory or class paradoxes springs from the fact that every truth-value statement can be rephrased as a statement about sets, and vice versa. For example, "All apples are red" means that the set of all apples is a subset of the set of all red things. This can be rephrased in truth-value language as the semantic statement: "If it is true that x is an apple, then it is true that x is red."

Consider the liar-paradox assertion: "This statement is false." It can be translated into the following set statement: "This assertion is a member of the set of all false assertions." If the statement actually does belong to the set of all false assertions, then what it asserts is true and therefore it cannot belong to the set of false statements. And if the statement does not belong to the set of all false assertions, then what it asserts is false, and therefore it must belong to the set of all false statements. Every semantic paradox has its analog in set theory, and every set theory paradox has its semantic analog.

Metalanguages

Semantic paradoxes are resolved by introducing the device of metalanguages. Statements about the world, such as "apples are red" or "apples are blue," are made in an *object language*. Statements about truth values must be made in a *metalanguage*.

A: Statement B is false.
B: Apples are blue.

In this example, there can be no paradox because sentence A, assumed to be written in metalanguage, talks about the truth value of sentence B, which is written in object language.

How can we talk about the truth values of a metalanguage? We must go to a higher metalanguage. Each rung of this infinite ladder is a metalanguage to the rung below, and an object language to the rung above.

The concept of metalanguages was developed by the Polish mathematician Alfred Tarski. At the bottom rung of the ladder are statements about objects, such as "Mars has two moons." Words like *true* and *false* cannot occur in this language. To speak about the truth or falsity of sentences in this language we must employ a metalanguage, the next higher rung of the ladder. The metalanguage includes all of the object language, but it is a "richer" language because it can talk about the truth values of the object language. To use Tarski's favorite example: "Snow is white" is a statement in an object language. But "The statement 'Snow is white' is true" is a statement in a metalanguage.

Can we speak of the truth and falsity of metalanguage statements? Yes, but only by going up to the third rung of the ladder and speaking in a still higher metalanguage that refers to all the languages below it.

Every rung of the ladder is an object language to the rung immediately above it. Every rung, except the bottom one, is a metalanguage to the rung immediately below. The ladder extends upward as far as we like.

Examples of sentences on the first four rungs of the ladder are:

A. The sum of the interior angles of any triangle is 180 degrees.

B. Sentence A is true.

C. Sentence B is true.

D. Sentence C is true.

Language at level A simply states theorems about geometrical objects. A geometry text containing proofs of the theorems is written in a metalanguage at level B. Books about proof theory are written in a metalanguage at level C. Fortunately, mathematicians seldom need to go beyond C.

The theoretical infinity of the ladder is amusingly discussed by Lewis Carroll in an article, "What the Tortoise Said to Achilles." A reprint of the article appears in *The Magic of Lewis Carroll,* by John Fisher, and in *Gödel, Escher, Bach,* by Douglas Hofstadter.

Theory of Types

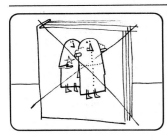

Set paradoxes are banished by a similar infinite hierarchy. A set cannot be a member of itself, or of any set of a lower type. The barber, astrologer, robot, and catalog simply don't exist.

The analog in set theory to Tarski's ladder of metalanguages is what Bertrand Russell originally called his "theory of types." Leaving technicalities aside, the theory arranges sets in a hierarchy of types in such a way that it is not permissible to say that a set is a member of itself, or not a member of itself. This eliminates self-contradictory sets. These potentially contradictory sets are simply ruled out of the system. There is no meaningful way to define them if you obey the rules of the theory of types. This corresponds to the semantic assertion that a sentence such as the liar paradox is simply "not a sentence" because it violates the formation rules of legitimate sentences.

Bertrand Russell spent many years working on his theory of types. In his book *My Philosophical Development,* Russell writes:

> When *The Principles of Mathematics* was finished, I settled down to a resolute attempt to find a solution of the paradoxes. I felt this as almost a personal challenge and I would, if necessary, have spent the whole rest of my life on an attempt to meet it. But for two reasons I found this exceedingly disagreeable. In the first place, the whole problem struck me as trivial. . . . In the second place, try as I would, I could make no progress. Throughout 1903 and 1904, my work was almost wholly devoted to this matter, but without any vestige of success.

The Swami's Prediction

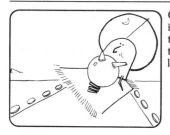

Can a swami see the future in his crystal ball? Predictions of the future can lead to a strange new kind of logic paradox.

The swami wrote on the card. At 3 o'clock Sue took the paper from under the crystal ball and read it aloud: "Before 3 P.M. you will write NO on the card."

One day the Swami had an argument with his teenage daughter Sue.
Sue: You're a big put-on, Dad. You can't really tell the future.
Swami: I most certainly can.
Sue: No you can't. And I can prove it!

Swami: You've tricked me. I wrote YES, so I was wrong. But if I'd written NO, it would be wrong too. There's *no* way I could have been right.
Sue: I'd like a red sports car, Dad, with bucket seats.

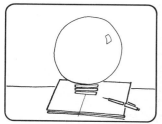

Sue wrote on a piece of paper, folded it, and stuck it under the crystal ball.
Sue: I've described an event that will either happen or not happen before 3 o'clock. If you can predict which it will be, you won't have to buy the car you promised me for graduation.

Sue: Here's a blank card. If you think the event will happen, write YES on it. If you think it won't, write NO. And if you're wrong, will you agree to buy me the car now instead of later?
Swami: Okay, Sue, it's a deal.

The original form of this paradox involves a computer that can only respond "yes" or "no." The computer is asked to predict whether its next response will be "no." Clearly, it is logically impossible for the prediction to be correct. The paradox can be reduced to ultimate simplicity by saying to someone: "Will the next word you speak be 'no'? Please answer 'yes' or 'no'."

Is this the same as the liar paradox? What is the meaning of "no" when the person replies? Clearly, it means "It is false that I am now saying 'It is false'." This in turn is the same as "This sentence is false." Thus the swami prediction paradox is little more than a disguised version of the liar paradox.

Note that just as "This sentence is true" does not lead to paradox, neither does the question, "Will the next word you speak be 'yes'?" The person can answer either "yes" or "no" without contradiction. As in the crocodile version of the liar paradox, this corresponds to the fact that the crocodile can either eat or return the baby, without contradiction, if the mother says: "You will return my baby."

The Unexpected Tiger

Princess: You're the king, father. May I marry Michael?
King: My dear, you may if Mike kills the tiger behind one of these five doors. Mike must open the doors in order, starting at 1. He won't know what room the tiger's in until he opens the right door. It will be an *unexpected* tiger.

When Mike saw the doors he said to himself:
Mike: If I open four empty rooms I'll *know* the tiger's in room 5. But the king said I wouldn't know in advance. So the tiger *can't* be in room 5.

Mike: Five is out, so the tiger must be in one of the other four rooms. What happens after I open three empty rooms? The tiger will have to be in room 4. But then it won't be unexpected. So 4 is out too.

By the same reasoning, Mike proved the tiger couldn't be in room 3, or 2, or 1. Mike was overjoyed.
Mike: There's no tiger behind *any* door. If there were, it wouldn't be unexpected, as the king promised. And the king *always* keeps his word.

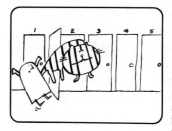

Having proved there was no tiger, Mike boldly started to open the doors. To his surprise, the tiger leaped from room 2. It was completely unexpected. The king had kept his word. So far logicians have been unable to agree on what is wrong with Mike's reasoning.

The paradox of the unexpected tiger has many other story forms. Of unknown origin, it first appeared in the early 1940s as a paradox about a professor who announced that an "unexpected examination" would be given on one day of the following week. He assured his students that no one could deduce the day of the examination until the day it occurred. A student "proved" it couldn't be on the last day of the week, or the next-to-last, or the day before that, and so on for all days of the week. Nevertheless, the professor was able to keep his word by giving the examination on, say, the third day.

When the Harvard University philosopher W. V. Quine wrote a paper about the paradox in 1953, it took the form of a warden who scheduled an unexpected hanging for a prisoner. For a discussion of the paradox, and a bibliography of 23 references, see the first chapter of my book, *The Unexpected Hanging and Other Mathematical Diversions.*

Most people admit that the first step in Mike's reasoning is correct, namely that the tiger cannot be in the last room. But once this is admitted as a sound deduction, the rest of Mike's reasoning seems to follow. For if the tiger cannot be in the last room, then identical reasoning rules out the next-to-last, and so on for the others.

However, even the first step of Mike's reasoning is faulty. Suppose he has opened all doors but the last. Can he deduce correctly that there is no tiger in the last room? No, because if he makes such a deduction, he might open the door and find an unexpected tiger! Indeed, the entire paradox holds even if only one room is involved.

Suppose Mr. Smith, who you believe always speaks truly, hands you a box and says, "Open it and inside you will find an unexpected egg." What can you deduce about the presence or absence of an egg in the box? If Smith is correct, the box must contain an egg, but then you will expect the egg and therefore Smith's statement is false. On the other hand, if this contradiction prompts you to deduce that the box cannot contain an egg (in which case Smith spoke falsely) and you open it to find an unexpected egg, then Smith spoke truly.

The consensus among logicians is that although the king knows he can keep his word, there is no way that Mike can know it. Therefore, there is no way he can make a valid deduction about the absence of the tiger in any room, including the last one.

Newcomb's Paradox

One day Omega, a superbeing from outer space, landed on the earth.

Omega had advanced equipment for studying human brains. He could predict with great accuracy how any person would choose between two alternatives.

Omega tested many people by using two large boxes. Box A was transparent and always held $1000. Box B was opaque. Either it was empty or it held 1 million dollars.

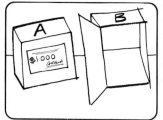

Omega told each subject:
Omega: You have two choices. One is to take both boxes and keep their contents. But if I expected you to do this, I have left B empty. You get only $1000.

Omega: Your other choice is to take only box B. If I expected you to do this, I have put a million dollars in B. You get it all.

This man has decided to take only box B. He reasons:
Man: I've watched Om make hundreds of tests. Every time he predicted right. Each person who took both boxes got only a thousand. So I'll take only box B and become a millionaire.

This woman has decided to take both boxes. She reasons:
Woman: Om has already made his prediction and left. Box B is not going to change. If empty, it stays empty. If full, it stays full. So I'll take both boxes and get everything that's here.

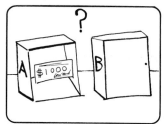

Who do you think made the best decision? Both arguments can't be correct. Which is wrong? *Why* is it wrong? This is a new paradox, and experts do not yet know how to solve it.

This is the latest and most bewildering of the many prediction paradoxes philosophers are currently debating. It was invented by a physicist, William Newcomb, and is known as Newcomb's paradox. A Harvard University philosopher, Robert Nozick, was the first to publish and analyze it. His analysis draws heavily on what mathematicians call "game theory" and "decision theory."

The man's decision to take only box B is easy to understand. To make the woman's argument clearer, recall that Omega has gone. Box B is either full or empty, and it is not going to change. If full, it remains full. If empty, it remains empty. Let's consider the two cases.

If B is full, and the woman takes only B, she gets a million dollars. But if she takes both boxes she gets a million plus a thousand.

If B is empty, and she takes only B, she gets nothing. But if she takes both boxes, she gets at least a thousand.

In each case, therefore, the woman is richer by a thousand dollars if she takes both boxes.

The paradox is a sort of litmus paper test of whether a person does or does not believe in free will. Reactions to the paradox are almost equally divided between believers in free will, who favor taking both boxes, and believers in determinism who favor taking only box B. Others argue that conditions demanded by the paradox are contradictory regardless of whether the future is or is not completely determined.

For a discussion of these conflicting views, see my Mathematical Games Department in *Scientific American*, July 1973, and the guest column by Professor Nozick in the same department, March 1974.

0   ZERO

1   ONE

2   TWO

3   THREE

4   FOUR

5   FIVE

6   SIX

7   SEVEN

8   EIGHT

9   NINE

Paradoxes about integers, fractions, and an infinite ladder

The history of mathematics has been strongly influenced by number paradoxes that have startled and confounded mathematicians by violating our intuition. Classic instances are the discoveries of:

1. Irrational numbers: $\sqrt{2}$, π, e, and an uncountable infinity of others.

2. Imaginary numbers: $\sqrt{-1}$ and the complex number system of which the imaginaries are part.

3. Numbers, such as quaternions, that violate the commutative law of multiplication, $a \times b = b \times a$.

4. Numbers, such as Cayley numbers, that violate the associative law of multiplication, $a \times (b \times c) = (a \times b) \times c$.

5. Transfinite or infinite numbers, such as the aleph numbers discovered by Georg Cantor, which opened up what David Hilbert, the great German mathematician, called a new "paradise" for mathematicians.

The paradoxes in this chapter are about rational numbers except for the last three, which contain irrational numbers and transfinite numbers. They have been selected not only to amuse you, but also to invite you to explore on your own some of the more significant regions of number theory into which they lead. For example, "The Ubiquitous Number 9" leads into finite arithmetics. "The Curious Will" leads into Diophantine analysis. Many of the paradoxes are jump-off points for generalized algebraic solutions that will polish your algebraic skills. The chapter closes with a tantalizing glimpse into Cantor's paradise, a field in which much exciting research is now going on.

The Six-Chair Mystery

Six students made reservations at a popular discotheque. At the last minute, a seventh student joins the group.

Hostess: Thank heavens, those kids are finally here! I've been holding six seats for them. Oh no! I see *seven!*

Hostess: No problem, though. I'll just seat the first student and let him hold his girlfriend on his lap for a few minutes.

Hostess: Now the third student sits next to the first two, and the fourth student sits next to her. Then the fifth one goes opposite the boy with the girl on his lap, and the sixth sits next to him. That takes care of six students, and there's *still* a vacant chair!

Hostess: So—all I have to do is tell the seventh student to get off her boyfriend's lap, walk around the table, and sit in the vacant chair!

Isn't that something? Seven persons seated in six chairs, one to a chair!

You should have no difficulty spotting the fallacy in this version of an old paradox about the landlady who puts 21 guests in 20 rooms. The paradox is resolved by realizing that the girl who sits temporarily on the boy's lap is, in fact, number 2. By the time the sixth student is seated, the discotheque owner has forgotten the girl's number and counts her as number 7. The actual seventh student does not get to the table at all. Number 2 simply gets off the boy's lap, moves around the table, and sits in the sixth chair.

The paradox appears to violate the theorem that a finite set of n elements can be put in one-to-one correspondence only with other sets that also have n elements. We return to this theorem when we consider infinite sets in the "Hotel Infinity" paradox. "The Six-Chair Mystery" is an amusing way to illustrate the difference between finite and infinite sets.

The Elusive Profit

Dennis sold one of his paintings to George for $100.
Dennis: You've got a bargain, George. In ten years that picture will be worth ten times as much.

George figures it differently.
George: By George, that artist sold his picture for $100 and bought it back for $80. That's a clear profit of $20. We can forget the next sale because $90 is about what the picture is worth.

George hung the painting in his home, but later he decided he didn't like it. He sold it back to Dennis for $80.

Gerry accepts both arguments.
Gerry: The artist made $20 when he sold his picture for $100 and bought it back for $80. Then he made another $10 when he paid $80 for it and sold it to me for $90. So his total profit is $30. What is the real profit? $10? $20? $30?

A week later, Dennis sold the picture to Gerry for $90.
Dennis: You've got a great bargain, Gerry. In ten years that picture will be worth *fifty* times what you paid for it!

The artist was pleased.
Dennis: First I sold the picture for $100. That just covered my time and materials, so it was an even trade. Then I bought it back for $80 and sold it for $90, so I'm $10 ahead.

This confusing little puzzle always provokes lively arguments. It may take some time to realize that the difficulty with this problem is that it is not "well defined," and that one answer is as good (or bad) as any of the other answers.

It is impossible to say what the artist's "real profit" is because the statement of the problem does not establish the initial "cost" of the painting. Put aside the cost of the artist's time spent on making the picture and say that Dennis paid a total of $20 for all the materials used, such as the frame, canvas, and paint. At the end of the three sales, the artist has obtained $110. If we define the final profit as the difference between the cost of his materials and the amount of money he ultimately received, then his profit is $90.

Since we do not know what the materials cost (we only assigned a value), we have no way of calculating the real profit. This problem *seems* to be arithmetical, but actually it is a debate over what is meant by real profit. This paradox is like the old question about whether a tree falling in a forest makes a sound if there are no ears to hear it. The answer can be yes or no depending on what is meant by the word *sound*.

The first two paradoxes in Chapter 3, Geometry, provide two other entertaining examples of problems that are basically arguments over what is being referred to by a word.

Population Implosion

We hear a lot these days about how fast the earth's population is growing.

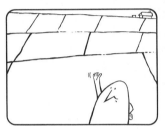

Mr. Ninny, president of the League Against Birth Control, disagrees. He thinks the world's population is *decreasing,* and soon everyone will have *more* space than he or she needs. Here's his argument.

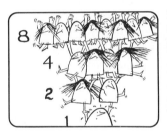

Mr. Ninny: Every person alive has two parents. Each parent had two parents. That makes four grandparents. And each grandparent had two parents, so that makes eight great-grandparents. The number of ancestors *doubles* for each generation you go back.

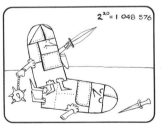

$2^{20} = 1\,048\,576$

Mr. Ninny: If you go back 20 generations to the Middle Ages, you would have 1,048,576 ancestors! And this applies to *every* person alive today. So the population of the Middle Ages must have been a million times what it is now! Mr. Ninny can't be right, but where's the flaw in his reasoning?

Ninny's argument is correct if the following two assumptions are made:

1. On the ancestral tree of every living person, no ancestor appears more than once.

2. The same person never appears on more than one tree.

Neither assumption can be correct in all cases. If a couple has five children, and each of these children has five children, the original couple will be grandparents on 25 separate trees. Moreover, on any one tree, if you go back many generations, there will be an overlap of branches arising from the marriage of distant relatives.

The fallacy of Ninny's argument is that it does not take into account either the duplications on single trees or the enormous "intersection" of the sets of people that make up each living person's tree. In Ninny's implosion argument, millions of people are counted millions of times!

Most people are surprised at how rapidly the terms of a doubling series increase. If someone agrees to give another person $1 today, $2 tomorrow, $4 the next day, and so on, it is hard to believe that on the 20th day the donor will be giving more than a million dollars!

Does a shortcut exist for obtaining the sum of the first 20 terms in this doubling series? Yes; double the last term, then subtract 1. The twentieth term is 1,048,576. The sum of the first 20 terms is

$$(2 \times 1,048,576) - 1 = 2,097,151$$

The trick applies to any partial sum of the terms in this doubling series. There is a simple way of showing this rule always works. Discovering that simple way is a challenging puzzle you might wish to try.

The Ubiquitous Number 9

The number 9 has many mysterious properties. Did you know that 9 is hidden inside the birthdate of every famous person?

Take George Washington's birthday. He was born February 22, 1732. Write that as the single number 2221732. Now rearrange these digits to make any different number. Subtract the smaller from the larger.

Add all the digits in the difference. In this case the sum is 36. And 3 plus 6 is 9!

If you do this with John Kennedy's birthdate (May 29, 1917), or Charles De Gaulle's (November 22, 1890), or that of any famous man or woman, you always get 9. Is there some curious connection between 9 and the birthdates of famous people? Does this work with *your* birthdate?

If all the digits of a number are added and then the digits of this sum are added, and this procedure is continued until only one digit remains, this final digit is called the *digital root* of the original number. The digital root is equal to the remainder when the original number is divided by 9, and for this reason the procedure is often called "casting out nines."

The fastest method of obtaining a digital root is to cast out nines while adding the original number's digits. For example, if the first two digits are 6 and 8, which add to 14, add 1 and 4 at once, and remember only 5. In other words, whenever a partial sum is more than one digit, add the two digits and carry only the sum. The final digit will be the digital root. The digital root is said to be equivalent to the original number modulo 9, usually abbreviated mod 9. Since 9 divided by 9 has a remainder of 0, in mod 9 arithmetic 9 and 0 are equivalent.

Before calculating machines existed, accountants often used mod 9 arithmetic to check the addition, subtraction, multiplication, and division of large numbers. If, for instance, we take A from B to get C, the work can be checked by taking the digital root of A from the digital root of B, and then seeing if the result matches the digital root of C. If the original subtraction is correct, there will be a match. This does not prove that the original subtraction is correct, but if there is *no* match, the accountant knows he or she has made a mistake. If there is a match, the result is probably correct. Similar digital-root checks apply to addition, multiplication, and division.

Now we are in a position to understand why the birthdate trick works. Suppose a number N is composed of many digits. We may scramble the digits to get a new number N'. Clearly, N and N' have the same digital roots. Therefore, if we subtract one digital root from the other, the difference will be 0, which is the same as 9 (in mod 9 arithmetic). This number, 0 or 9, must be the digital root of the difference between N and N'. In short, take any number whatever, scramble its digits, subtract the smaller from the larger, and the difference will have a digital root of 0 or 9.

Because of the way the digital root is calculated, a final result of 0 can occur only if N and N' are identical numbers. Thus, when they try the procedure on their birthdates, your friends should be sure that scrambling produces a different number. As long as the two numbers are not the same, the difference will have a digital root of 9.

Many magic tricks are based on the ubiquitous number 9. For example, ask someone to write down the number of a dollar bill while your back is turned so that you cannot see what the person is writing. The person then scrambles the digits to make a different number and takes the smaller from the larger. Ask your friend to cross out any single nonzero digit in the result, then read the remaining digits aloud *in any order*. With your back still turned, you should have no trouble naming the crossed-out digit!

The secret of this trick should be obvious. The difference will have a digital root of 9. As your friend calls out the digits, add them mentally, casting out nines. When the person is finished, subtract the final digit from 9, and the digit crossed out is the result. (If the final digit is 9, then the crossed-out number is 9.)

The birthdate and dollar bill tricks are excellent introductions to the study of modular arithmetic systems.

The Bewildered Bus Driver

This bus is filled with 40 boys. Soon they will be on their way to camp.

This bus is filled with 40 girls. They are going to the same camp.

Before starting, the bus drivers have some coffee.

Meanwhile, ten boys get off their bus and sneak into the girls' bus.

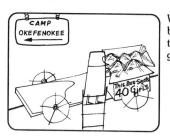
When the driver of the girls' bus comes back, he notices there are too many passengers.

Driver: All right, let's cut out the fun and games! This bus seats 40 people, so 10 of you had better get off. And make it fast!

Ten passengers of unknown sex get off. They all board the boys' bus and take the ten empty seats. Soon the two buses, each with 40 campers, are on their way.

Later, the driver of the girls' bus thinks:
Driver: Hmm . . . I'm sure there are some boys on this bus, and some girls on the boys' bus. I wonder which bus has the most passengers of the wrong sex?

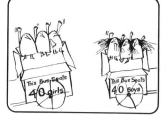
It's hard to believe, but regardless of the sexes of the 10 campers who go back to the boys' bus, the buses will have *exactly* the same proportion of the opposite sex.

Why? Suppose 4 boys are on the girls' bus. This leaves 4 empty seats on the boys' bus. These empty seats have to be filled by four girls. And the same argument applies to any other number of boys.

This paradox can easily be demonstrated with a deck of playing cards. First, the deck is divided into 26 red and 26 black cards. Let someone cut a packet of cards from one of the piles. Let us say he takes 13 cards from the red pile. Placing them on top of the black pile, he thoroughly shuffles the pile. He is now instructed to remove the same number of cards (in this case 13) from the pile just shuffled (they can be taken randomly from anywhere in the pile) and to place them on the red pile. Finally, this restored half-deck is also shuffled.

When the two half-decks are examined, you will find that the number of red cards in the black packet exactly matches the number of black cards in the red packet. The proof of this stunt is identical to that for the numbers of boys and girls on the two buses.

Many card tricks are based on this principle. Here is another one in which the principle is nicely concealed. Divide the deck exactly in half, turn one half face up, and shuffle the two halves together. Show this mixed-up deck to your audience without saying that exactly 26 cards have been turned face up. Allow someone to shuffle the deck thoroughly. Hold out your hand and ask him to deal 26 cards on your palm.

"Wouldn't it be an astonishing coincidence," you say, "if my half contains precisely the same number of face-up cards as your half?"

Ask him to spread his half on the desk top. As he does so, secretly *turn over your half* before you spread it on the desk beside his half. Count the number of face-up cards in each half. The two numbers will be the same!

The trick is based on the same principle involved in the bus paradox. If you had not turned over your half, the number of face-up cards in the other half would exactly match the face-down cards in your half. When you turn over your half,

your face-down cards become face-up, and this puts them in one-to-one correspondence with the face-up cards in the other half.

At this point, we may consider the following old brainteaser. A glass of water stands next to a glass of wine. The two quantities of liquid are equal. A drop of wine is transferred to the glass of water. The water is thoroughly stirred, then a single drop (the same size as the previous one) is taken from the mixture and put back in the glass of wine. Is there now more or less wine in the water than there is water in the wine?

The quantities of the two mixtures will be the same. The answer does not change even if the glasses hold different quantities of liquid, or whether the mixture is thoroughly stirred. Moreover, we may transfer drops of different sizes back and forth as often as we like. The only condition that must be met is that, at the finish, each glass must hold the same amount of liquid as it held at the beginning. The wine glass, for instance, will then be missing a certain amount of wine. The place of this missing wine will be filled by exactly the same amount of water! The proof of equality for this brainteaser is the same as the proof for the number of boys and girls on the two buses or the number of red and black cards in the two half decks.

The wine and water problem is a marvelous example of a problem that can be solved by a tedious algebraic proof, but which yields readily to a simple logical proof if one only has the right insight.

The Missing Dollar

A record store put 30 old rock records on sale at two for a dollar, and another 30 on sale at three for a dollar. All 60 were gone by the end of the day.

What do you think happened to that missing dollar? Did the clerk steal it? Did a customer get the wrong change?

The 30 two-for-a-dollar disks brought in $15. The 30 three-for-a-dollar disks brought in $10. Altogether—$25.

The next day the store manager put another 60 records on the counter.
Clerk: Why bother to sort them? If 30 sell at two for a dollar, and 30 at three for a dollar, why not put all 60 in one pile and sell that at five for $2? It's the same thing.

When the store closed, all 60 records had been sold at five for $2. But when the manager checked the cash, he was surprised to find that proceeds from the sale were only $24, not $25.

Let's figure out just what is going on here. As the story shows, the clerk is mistaken in his hunch that selling two sets of records at five for $2 is "the same thing" as selling them in separate stacks at two for a dollar and three for a dollar. There is no reason why the money taken in should be the same in both these cases. In this case the difference is so small—only a dollar—that it seems as if a dollar may have been overlooked or even lost.

Consider the same problem but with slightly different parameters. Suppose the more expensive set of records sell at three for $2, or a price of 2/3 dollar per record. The less expensive records are two for $1, or 1/2 dollar per record. The manager combines the two stacks and sells them at five for $3. If there are 30 records in each set, as before, selling the sets separately brings in $35, but selling all 60 at the combined price brings in $36. Now the store has made an extra dollar rather than lost one!

There is nothing wrong in trying out the clerk's hunch, but the numbers show he was wrong. The error can be analyzed algebraically, but a more extreme example is sufficient to show that you cannot average prices and numbers of units per transaction, in the manner described, and get equivalent results.

Consider a car dealer who has six Rolls Royces and six Volkswagens. He puts them on sale at two Rolls for $100,000 and six Volks for $50,000. If he sells all twelve cars he gets $350,000. The average number of cars per single sale of each type is four. The average of the two transaction prices is $75,000. Now if he were to put his entire stock on sale at four cars for $75,000, and sold all the cars, he would get only $225,000. Moreover, a customer would surely buy four Rolls for $75,000, leaving the dealer with a stock of eight overpriced cars. So much for the clerk's reasoning in the original problem!

Magic Matrix

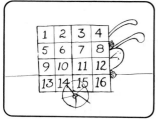

Copy this 4-by-4 matrix on a sheet of paper and number the cells from 1 to 16. I'm going to astound you with a remarkable demonstration of psychic power! I'm going to control your selection of four numbers on this matrix.

Are you ready? I'm going to tell each of you what your total is. It . . . is . . . 34! Right? How do I know? Was I really able to influence your selection?

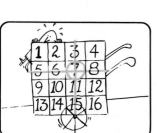

Draw a circle around any number you like. This picture shows a circled 7, but you may pick any number you please. Now draw a vertical line through the column that contains your number. Then draw a horizontal line through the row that contains your number.

Circle any number not already crossed out. Again, draw lines through the number's column and row. Choose a third number in the same way, and cross out its row and column. Finally, circle the single number that remains.

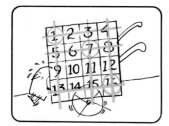

If you followed directions, your square will look something like this. Now, add the four numbers you selected.

Why does the matrix force us always to choose four numbers that add up to 34? The secret is ingenious and simple. At the top of each column of a 4-by-4 matrix, place the numbers 1, 2, 3, 4. At the left of each row put the numbers 0, 4, 8, 12.

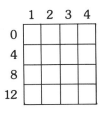

These eight numbers are called the *generators* of the magic matrix. Each cell is now given a number that is the sum of its two generators, the generator beside its row and the generator above its column. When we have filled in all the cells, we will have a matrix bearing the numbers 1 to 16 in counting order.

	1	2	3	4
0	1	2	3	4
4	5	6	7	8
8	9	10	11	12
12	13	14	15	16

Now let us see what happens when four numbers are circled in accordance with the procedure described. The procedure guarantees that *no two circled numbers will be in the same row or column*. Each number is the sum of a unique pair of generators, therefore the sum of the four circled numbers will equal the sum of the eight generators. Since the eight generators add to 34, the four circled numbers must also add to 34.

When you understand how the matrix works, you should be able to make a magic matrix of any size. Consider, for example, the order-6 matrix below, with its 12 generators. Notice that in this case the generators were chosen so that the

numbers in the cells appear to be random. This conceals the underlying structure of the matrix, and thus makes it *seem* more mysterious.

	4	1	5	2	0	3
1	5	2	6	3	1	4
5	9	6	10	7	5	8
2	6	3	7	4	2	5
4	8	5	9	6	4	7
0	4	1	5	2	0	3
3	7	4	8	5	3	6

The generators sum to 30. If 6 numbers are chosen in accordance with the procedure, they will sum to 30. The forced number (or sum) can, of course, be any number we wish.

You can construct a 10-by-10 matrix that will force the number (or sum) 100, or any other interesting number such as the current year or a person's year of birth. Can magic matrices be constructed with negative numbers in some cells? Of course! In fact, a generator may be any number, positive or negative, rational or irrational.

Can a magic matrix be constructed in which the selected numbers are *multiplied* rather than added to get the final number? Yes; this suggests another path to explore. The basic construction is exactly the same. The forced number, in this case, becomes the *product* of the set of generators. You might also wish to investigate what happens if complex numbers are used in the cells. For more material on magic matrices, consult the second chapter of my *Scientific American Book of Mathematical Puzzles and Diversions.*

The Curious Will

A wealthy lawyer owned 11 antique cars, each worth about $25,000.

When the lawyer died, he left a curious will. It asked that his 11 cars be divided among his three sons. Half of the cars were to go to the eldest son, a fourth to the middle son, and a sixth to the youngest.

Everybody was puzzled. How can 11 cars be divided into two equal parts? or four? or six?

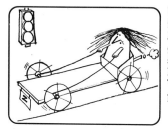

While the sons were arguing about what to do, Ms. Zero, the famous numerologist, drove up in her new sports car.
Ms. Zero: Hello, boys. You look as if you have a problem. Can I help?

After the sons explained the situation, Ms. Zero parked her sports car next to the 11 antique cars and hopped out.
Ms. Zero: Tell me, boys, how many cars are here? The boys counted 12.

Then Ms. Zero carried out the terms of the will. She gave half of the cars, or six, to the oldest son. The middle son got a fourth of 12, or three. The youngest son got a sixth of 12, or two.
Ms. Zero: 6 plus 3 plus 2 is just 11. So, one car is left over. And that's *my* car.

Ms. Zero hopped into her sports car and drove off.
Ms. Zero: Always glad to be of help, boys! I'll send you my bill!

This is a modern version of an old Arabian paradox involving horses rather than cars. You can vary the terms of the will by changing the number of cars and dividing them up by a different set of fractions, subject to the condition that the borrowing of one car permits carrying out the terms of the will with one car left over to be returned to the lender.

For example, there could be 17 cars and a will that says these cars are to be divided into halves, thirds, and ninths. If there are n cars, and the three fractions are $1/a$, $1/b$, and $1/c$, the paradox holds only if the equation

$$\frac{n}{n+1} = \frac{1}{a} + \frac{1}{b} + \frac{1}{c}$$

has a solution in positive integers. See if you can elaborate on the problem by increasing the number of heirs, as well as increasing the number of cars to be borrowed in order to carry out the will.

The resolution of the paradox lies, of course, in the fact that the fractions decreed by the original will have a sum that is less than 1. If the will were carried out by actually cutting up the cars, 11/12 of a car would be left over. Ms. Zero provides a way of distributing that 11/12 to the sons. Thus, the oldest gets 6/12 of a car more than he would have gotten before, the middle son gets 3/12 more, and the youngest son gets 2/12 more. These three fractions add to 11/12, and since each son now gets an integral number of cars, no cutting is necessary.

The Amazing Code

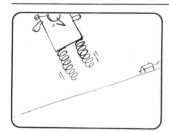

Dr. Zeta is a scientist from Helix, a galaxy in another space-time dimension. One day Dr. Zeta visited the earth to gather information about humans. His host was an American scientist named Herman.

Using his powerful pocket computer, Dr. Zeta scanned the encyclopedia quickly, translating its entire content into one gigantic number. By putting a decimal point in front of the number, he made it a decimal fraction.

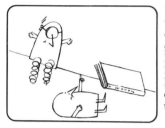

Herman: Why don't you take back a set of the Encyclopedia Britannica? It's a great summary of all our knowledge.
Dr. Zeta: Splendid idea, Herman. Unfortunately, I can't carry anything with that much mass.

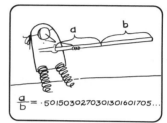

Dr. Zeta then placed a mark on his rod, dividing it accurately into lengths a and b so that the fraction a/b was equivalent to the decimal fraction of his code.

$$\frac{a}{b} = .50150302703013016017\overline{05}\ldots$$

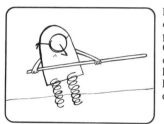

Dr. Zeta: However, I can encode the entire encyclopedia on this metal rod. One mark on the rod will do the trick.
Herman: Are you joking? How can one little mark carry so much information?

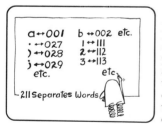

Dr. Zeta: When I get back to my planet, one of our computers will measure a and b exactly, then compute the fraction a/b. This decimal fraction will be decoded, and the computer will print your encyclopedia for us!

Dr. Zeta: Elementary, my dear Herman. There are less than a thousand different letters and symbols in your encyclopedia. I will assign a number from 1 through 999 to each letter or symbol, adding zeros on the left if needed so that each number used will have three digits.

Herman: I don't understand. How would you code the word *cat?*
Dr. Zeta: It's simple. We use the sort of code I just showed you. *Cat* might be coded 003001020.

If you are not already familiar with ciphers, you may enjoy coding and decoding some simple messages in a number code similar to the one used here. Codes illustrate the importance of one-to-one correspondence, and the mapping of one structure onto an isomorphic structure. Such codes are actually used in advanced proof theory. There is a famous proof by Kurt Gödel that every deductive system complicated enough to contain the integers has theorems that cannot be proved true or false within the system. Gödel's proof is based on a number code that translates every theorem of a deductive system into a unique and very large integer.

Coding an entire encyclopedia by placing one mark on a rod works only in theory, not in practice. The difficulty is that the precision needed for marking such a rod is impossible to achieve. The mark would have to be enormously smaller than an electron, and the measurements of the two lengths would have to be precise on the same scale. If we assume that two lengths can be measured accurately enough to yield Dr. Zeta's fraction, then of course his procedure *would* work.

Switching to **ir**rational numbers, mathematicians believe that the decimal expansion of π (pi) is as "unpatterned" as any typical infinite sequence of random digits. If this is true, it means that somewhere in the expansion, any finite sequence of digits is certain to appear. In other words, at some spot in the decimal expansion of π is a sequence that codes the Encyclopedia Britannica as Dr. Zeta did or, indeed, a sequence that codes any other work that has been printed or that *could* be printed!

There also are strongly patterned irrational numbers that contain every finite sequence of digits. An example is the number .123456789101112131415. . . , formed by writing the counting numbers in counting order.

Hotel Infinity

Before Dr. Zeta left, he told a fantastic story.

Dr. Zeta: Hotel Infinity is an enormous hotel at the center of our galaxy. It has an infinite number of rooms that extend through a black hole into a higher dimension. The room numbers start at 1 and go on forever.

Dr. Zeta: One day, when every room was occupied, a UFO pilot, on his way to another galaxy, arrived.

Dr. Zeta: Even though there was no vacancy, the hotel manager found a room for the pilot. He just moved the occupants of each room to the room with a number that was one higher. This left Room 1 vacant for the pilot.

Dr. Zeta: The next day, five couples on their honeymoons showed up. Could Hotel Infinity take care of them? Yes, the manager simply moved everybody to a room with a number that was five higher. This left rooms 1 through 5 vacant for the five couples.

Dr. Zeta: On the weekend an *infinite* number of bubble gum salespeople came to the hotel for a convention.

Herman: I can understand how Hotel Infinity could take care of any *finite* number of new arrivals. But how could it find room for an *infinite* number?

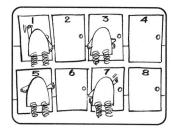

Dr. Zeta: Easily, my dear Herman. The manager just moved everyone to a room with a number *twice* as large as before.

Herman: Of course! That put everybody in a room with an *even* number. This left all the odd-number rooms—an infinity of them—vacant for the bubble gummers!

No finite set can be put into one-to-one correspondence with one of its proper subsets. This is not true of infinite sets. They seem to violate the old rule that a whole is greater than any of its proper parts. Indeed, an infinite set can be defined as one that can be put in one-to-one correspondence with a proper subset of itself.

The manager of Hotel Infinity first showed how the set of all counting numbers can be put in one-to-one correspondence with one of its proper subsets so as to leave one element left over, or five elements left over. Clearly, this procedure can be varied so that an infinite subset can be taken from the entire set, leaving any desired finite number of elements.

Another way to dramatize this kind of subtraction is to imagine two infinitely long measuring rods lying side by side on a desk, their zero ends flush and at the center of the desk. Both rods are marked and numbered in centimeters. They extend infinitely far to the right, with all numbers in one-to-one correspondence: 0–0, 1–1, 2–2, and so on. Now imagine sliding one rod n centimeters to the right. After this operation all marks on the rod that was moved will still be in one-to-one correspondence with marks on the stationary rod. If the rod were moved 3 centimeters, the marks will correspond as 0–3, 1–4, 2–5, ... The n centimeters that project represent a difference in lengths between the two rods. Both rods remain, however, infinitely long. Since we can make n, the difference, any value we please, it is clear that subtracting from infinity is an ambiguous operation.

The hotel manager's final maneuver opened up an infinite number of rooms. This shows how infinity can be taken from infinity yet leave infinity. By putting every counting number in one-to-one correspondence with every even counting number, an infinite set of whole numbers— namely all the odd ones—is left over.

The Ladder of Alephs

Hotel Infinity is only one of many paradoxes about infinite numbers. There are many infinities! The number of counting numbers is only the lowest infinity in an endless hierarchy. The second infinite number is the number of points in the entire universe, and the third infinite number is very much larger than that!

The German mathematician Georg Cantor, who discovered this ladder of infinities, called his strange new numbers aleph-null, aleph-one, aleph-two, and so on.

The *cardinal number* of a set is the number of the elements in the set. For example, the cardinal number of the set containing the letters of the word *cat* is 3. Any finite set has a finite cardinal number. Georg Cantor discovered that some infinite sets were "larger" than other infinite sets. He used the first letter in the Hebrew alphabet, aleph ($\aleph$), to denote the cardinal number of an infinite set. Subscripts specify which "infinity."

The cardinal number of the set of counting numbers Cantor called $\aleph_0$ (aleph-null). The set of even numbers and the set of odd integers both have cardinal number $\aleph_0$. Thus $\aleph_0 + \aleph_0 = \aleph_0$. The Hotel Infinity paradox showed that in some sense we can have $\aleph_0 - \aleph_0 = \aleph_0$! What crazy numbers!

The set of real numbers forms a larger infinite set, which Cantor believed to have cardinal number $\aleph_1$ (aleph-one), the first transfinite cardinal number greater than $\aleph_0$. His famous "diagonal proof" showed that the set of real numbers cannot be put in one-to-one correspondence with the set of integers. He also showed that the set of real numbers corresponds to the number of points on a line segment, on an infinite line, on a square, on an infinite plane, in a cube, in an infinite space, and so on for hypercubes and higher spaces.

When 2 is raised to the power of an aleph, Cantor proved that it generates a higher aleph that cannot be put in one-to-one correspondence with the aleph in the exponent. Thus the ladder of alephs continues upward forever.

The cardinality of the set of real numbers is known as c, or the "power of the continuum." Try as he would, Cantor was unable to prove that c is the same as aleph-one. Many decades later the work of Kurt Gödel and of Paul Cohen established that this question cannot be decided by using the axioms of standard set theory. As a result, set theory is now divided into Cantorian and non-Cantorian branches. Cantorian set theory assumes that $c = \aleph_1$. Non-Cantorian set theory assumes an infinity of transfinite numbers between $\aleph_0$ and c.

The famous "continuum hypothesis," as Cantor's conjecture came to be known, was resolved by showing it to be *undecidable*. The situation is similar to what happened after the discovery that Euclid's parallel postulate could not be proved. The postulate could be replaced by other possibilities, thus dividing geometry into Euclidean and non-Euclidean geometries.

Paradoxes about plane, solid, and impossible shapes

To most people the word *geometry* means Euclidean plane geometry—the study of properties of rigid plane figures. In this chapter we take the word in the broader sense proposed more than a century ago by Felix Klein. It is the study of properties of figures, in a space of any dimensions, that are invariant with respect to any defined group of transformations.

Klein's concept of geometry is one of the most seminal and unifying concepts of modern mathematics. In Euclidean plane and solid geometry the transformations that are permitted consist of translations (moving from one place to another), mirror reflections, rotations, and dilations (magnifying or diminishing). More extreme transformations define affine geometry, projective geometry, topology, and finally set theory, in which a figure can be broken up into points that may be rearranged.

According to Jean Piaget, the Swiss psychologist, children actually learn to grasp geometrical properties in reverse of the above order! Very young children, for instance, find it easier to distinguish between a pile of red marbles and a pile of blue marbles (set theory), or between a closed rubber band and one cut open (topology), than to distinguish between a pentagon and a hexagon (Euclidean geometry).

Topology is a strange branch of geometry that studies properties invariant with respect to continuous deformations. Think of an object as made of rubber which can be twisted and distorted any way you like provided you don't break off parts and stick them back again. One-sidedness, for example, is a topological property of a Moebius strip because, if you imagine it made of rubber, no amount of twisting and stretching can alter its one-sided property. Many paradoxes in this chapter—braiding a bracelet, turning a torus inside out, a fixed-point theorem, and others—deal with topological properties.

The reflection transformation, in which an asymmetric figure like the capital letter B is changed to its mirror image, is emphasized in this chapter not only because it underlies so many fascinating paradoxes, but also because it is so important in modern geometry and modern science. Mirror symmetry plays a fundamental role in chemistry, especially organic chemistry in which almost all carbon molecules are asymmetric, with left-handed and right-handed forms. It is also of major importance in crystallography, in biology and genetics, and in particle physics.

Although some of the paradoxes may seem at first to be little more than recreational curiosities, you will see that each can lead you smoothly into significant areas of mathematics such as group theory, logic, sequences, infinite series, and limits. Too often students of geometry become so concerned with ruler and compass constructions and the step-by-step proving of theorems that they miss the exciting relationships between geometry and other branches of mathematics, and the endless and beautiful applications of geometry to astronomy, physics, and other sciences.

Getting Around a Girl

Marvin: Oh Myrtle! Are you hiding behind that tree?

As Marvin circled the tree, Myrtle did the same thing. By keeping her nose against the tree as she sidled around it, she kept out of sight.

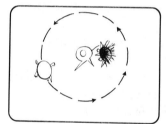

After going once around the tree, they were back where they started. Has the boy gone around Myrtle?
Marvin: Of course! I walked around the tree so I must have gone around *her*.
Myrtle: Nonsense! Even if the tree weren't there, he'd never see my back. How can you go *around* anything without seeing all sides?

This ancient paradox is usually given in the form of a hunter and a squirrel. The squirrel is sitting on a stump. As the hunter circles the stump, the squirrel keeps turning so that the squirrel is always facing the hunter. After the hunter has circled the stump, has he gone around the squirrel?

The question cannot be answered, of course, unless there is agreement on what is meant by the word *around*. Many words that are part of everyday speech do not have precise definitions. William James, in his classic philosophic work, *Pragmatism*, has an amusing discussion of the hunter and squirrel paradox. He presents it as a model of a disagreement that is purely semantic. The difficulties vanish as soon as the disagreeing parties realize they are only arguing over how to define a word. If people were only more aware of the importance of precise definitions of terms, many bitter arguments would turn out to be almost as trivial as this one.

The Great Moon Mystery

The moon always keeps the same face turned toward the earth. After it has made one revolution around the earth, has it rotated on its own axis?

Father: As an astronomer, I say yes. If you watched from Mars, you'd see the moon turn once on its axis each time it circled the earth.

Daughter: How *can* it rotate, father? If it did, we'd see different sides, but all we ever see is the same old side.
Does the moon rotate? Did the boy go around the girl? Are these genuine paradoxes, or just arguments over the meaning of a word?

Like the previous paradox, this one is another example of a semantic argument. What exactly is meant by the phrase *rotated on its own axis?* Relative to an observer on the earth, the moon does not seem to rotate. Relative to an observer outside the earth–moon system, it does.

It is hard to believe, but intelligent persons have taken this simple paradox with utmost seriousness. Augustus De Morgan, in the first volume of his *Budget of Paradoxes,* reviews several nineteenth century pamphlets attacking the notion that the moon rotates. Henry Perigal, a London amateur astronomer, was indefatigable in his arguments. According to an obituary, his "main astronomical aim in life" was to convince others that the moon does not rotate. Perigal wrote booklets, built models, and even composed poems to prove his point, "bearing with heroic cheerfulness the continual disappointment of finding none of them of any avail."

A marvelous little paradox, closely related to the moon question, can be discussed at this point. Draw two touching circles of equal size to represent two disks. One disk is to be rolled around the other without slipping, keeping the rims in contact. How many times will the rolling disk have rotated after it has completed one revolution around the fixed disk?

Most people will answer one. Let them try it with two coins of the same size, and they may be surprised to discover that the rolling coin actually rotates *twice!*

Or does it? As in the earth–moon paradox, it depends on the frame of reference of the observer. Relative to its initial point of contact with the fixed coin, the revolving coin rotates once. Relative to you, looking down on the coins, it rotates twice. This, too, has been the topic of furious controversy. When *Scientific American* published the problem originally in 1867, it produced a flood of letters from readers who took sharply opposing sides.

Readers were quick to perceive the relation of the coin paradox to the earth–moon paradox. Those who argued that the rolling coin rotates only once also argued that the moon does not rotate at all. "If you swing a cat around your head," wrote one reader, "would his head, eyes and vertebrae each revolve on its own axis. . . ? Would he die at the ninth turn?"

The volume of mail swelled to such proportions that in April 1868, the editors announced they were dropping the topic but would continue it in a new monthly magazine *The Wheel* devoted entirely to the "great question." At least one issue of this magazine appeared, featuring pictures of elaborate devices that readers had made and sent to the editors to prove their case.

The rotation of astronomical bodies sets up inertial effects, which can be detected by such devices as the Foucault pendulum. Such a pendulum on the moon would show that the moon does indeed rotate as it swings around the earth. Does this change the argument to one independent of the observer's frame of reference?

Surprisingly, in light of general relativity it does not. You can assume that the moon does not rotate at all, but that the entire universe (regardless of whether its space-time structure is independent of the matter it contains) rotates around the moon. This rotating universe creates gravity fields that produce the same effects as the inertial fields generated by a rotating moon in a fixed cosmos! Of course, it is much more convenient to think of the universe as the fixed frame. But strictly speaking, the question of whether any object "really" rotates or is fixed is, in relativity theory, a meaningless question. Only the relative motion is "real."

Mirror Magic

Mirrors are puzzling. Timothy and Rebecca are guests at a party where everybody wears a name tag.

If you stand with your *side* to a mirror, your left–right axis is perpendicular to the glass. Now, your head stays up, your front stays front, and you are reversed from left to right.

Rebecca: What a strange mirror, Tim! Look—it reverses *my* name but doesn't change yours at all!

When you *face* a mirror, your head stays up, your left stays left, and you are reversed from front to back. Because your image's left hand is opposite where it would normally be if you stepped behind the glass and turned around, we *say* that the mirror has reversed left and right.

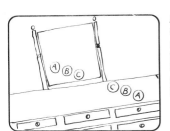

Isn't it mysterious that mirrors seem to reverse only left and right? Why doesn't a mirror also reverse up and down?

Why does this mirror reverse only CARBON and not DIOXIDE? It doesn't! The letters of DIOXIDE are reversed too, but their symmetry makes them look the same after they are reversed.

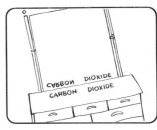

Actually, a mirror reverses only along lines perpendicular to its surface. Because these three balls are on a line at right angles to the glass, their order is reversed in the reflection.

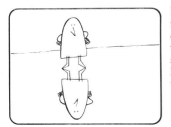

Can you guess what happens when two mirrors are placed at right angles? They create an *unreversed* image. Rebecca is seeing herself the way others *see* her!

If you stand on a mirror floor, your up–down coordinate axis is perpendicular to the glass. So your front stays front, your left stays left, and you are reversed from top to bottom.

Because each letter of TIMOTHY has a vertical axis of symmetry, the mirror image of the name appears unchanged. In REBECCA only the A has a vertical axis of symmetry. As a result, the A is unchanged, but all the other letters are mirror-reversed.

Why does a mirror reverse left and right, but not up and down? Similar to the paradoxes about the moon and the coins, this paradox also poses a semantic question that cannot be answered without agreement on the meanings of such words as *left, right,* and *reverse.* For a more detailed analysis of exactly what a mirror does, see the first three chapters of my *Ambidextrous Universe.* This book contains a great deal of material on mirror-reflection symmetry and the role it plays in science and everyday life.

Unlike the letters of TIMOTHY, those in DIOXIDE all have a *horizontal* axis of symmetry. Consequently, when a mirror is held *above* the word, all its letters appear unaltered in the reflection. In the word CARBON, C, B, and O appear the same in the mirror because they, too, have horizontal axes of symmetry. But A, R, and N, lacking such axes, are turned upside-down and are mirror-reversed.

What English words are unchanged by this kind of mirror reflection? The first step is to examine all capital letters and list those that have a horizontal axis of symmetry. They are B, C, D, E, H, I, K, O, X. With these letters we can form many words of four or more letters: CHOICE, COOKBOOK, ECHO, OBOE, ICEBOX, HIDE, DECIDED, CHOKED, and hundreds of others.

You can see an unreversed image of your face by holding two pocket mirrors at right angles and looking into the corner (the angle of the abutted mirrors must be adjusted until you see a single image of your face.) If you wink your left eye, your mirror image will not wink its right eye, as expected, but will wink the eye on the other side.

The two sides of your face have been switched because each side is reflected twice, once by each mirror.

Your face will probably look strange to you. That's because the face you see in ordinary mirrors is always reversed. Although faces have a vertical axis of symmetry, left and right sides are seldom perfect mirror images of each other. When you see your face unreflected, these slight differences between left and right sides cause your image to appear strangely different, even though you cannot say exactly why. Yet this is the face by which you are known to the world! What is more, the mirror image of your face looks equally strange to those who know you well.

A good way to check your understanding of how the double mirror works is to ask yourself what you will see if you turn the two mirrors so that the touching edges are horizontal rather than vertical. The double reflection will now turn your face upside down! Is this inverted face a mirror-reversed image? No, it is still unreversed. If you wink your left eye, your upside-down image will again wink the eye on the other side.

These mirror tricks are excellent introductions to the study of symmetry and reflections in transformation geometry. The paradoxes can all be explained by applying elementary transformation theory.

Cubes and Ladies

How many cubes do you count here? Are there 6? Are there 7?

Is this a drawing of a young woman? Or do you see an old hag?

What do you observe here? A small cube in the corner of a room? A small cube stuck on the *outside* of a large block? Or a large block with a cubical *hole* in one corner?

These optical illusions are all instances of a fluctuating interpretation of what you see. In the first illusion, your mind sees the flat pattern as a perspective drawing of a set of cubes; however, this drawing can be seen in two different ways. Each interpretation is equally good, so the mind switches back and forth between them.

The same description is true of the picture of the young woman or the old lady. It is impossible not to see one or the other, and the mind jumps back and forth between the two interpretations.

In the third illusion, there are *three* interpretations. For most people, seeing this illusion as a block with a cubical *hole* is the most difficult, because cubical holes in blocks are seldom seen. But if you keep looking and try to imagine the small cube as a hole instead of a solid, you will eventually see the picture that way. Learning to "see" this diagram in the three possible ways is closely related to the ability to interpret geometrical drawings. In geometry "seeing" a diagram incorrectly can be a major source of confusion.

Randi's Remarkable Rugs

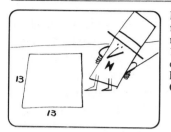

Mr. Randi, the world famous magician, owns a rug that is 13 decimeters by 13 decimeters. He wants to change it to an 8-by-21 rug. Mr. Randi took the rug to Omar, a rug dealer.

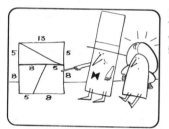

Randi: Omar, my friend, I want you to cut this rug into four pieces, then sew them together to make an 8-by-21 rug.
Omar: I'm sorry, Mr. Randi. You're a great magician but your arithmetic is terrible: 13-by-13 is 169, 8-by-21 is 168. It won't work.

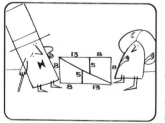

Randi: My dear Omar. The great Randi is *never* wrong. Kindly cut the rug into four pieces like this.

Omar did as he was told. Then Mr. Randi arranged the pieces, and Omar sewed them together to make an 8-by-21 rug.
Omar: I can't believe it! The area has shrunk from 169 to 168! What happened to that missing square decimeter?

This classic paradox is so startling and hard to explain that it is worth taking time to draw the square on graph paper, cut out the four pieces, and rearrange them to make the rectangle. Unless the pieces are very large, and drawn and cut with extreme precision, you will not notice the tiny overlap along the rectangle's main diagonal. It is this failure of the pieces to fit properly along the diagonal that accounts for the missing square unit of area. If you doubt the existence of this overlap, one way to prove it is to calculate the slope of the rectangle's diagonal and compare it with the slopes of the pieces.

What happens if the rectangle is drawn on the graph paper, the pieces cut, and then formed into the square? You might wish to investigate this.

Four lengths are involved in this paradox: 5, 8, 13, and 21. You may recognize these numbers as four terms in a famous sequence. Can you give the recursive rule for the terms? The sequence is the Fibonacci sequence in which each term is the sum of the two preceding terms: 1, 1, 2, 3, 5, 8, 13, 21, 34, ...

Variants of the paradox are based on other sets of four consecutive terms in the Fibonacci sequence. In every case you will find that the rectangle has a different area from the square, but sometimes the rectangle gains an extra square unit, sometimes it loses. The next step is the discovery that when there is a loss it is because of a rhombus-shaped overlap along the rectangle's diagonal, and where there is a gain it is because of a rhombus-shaped *gap*.

Given the four terms of the Fibonacci sequence on which a variant is based, can one predict whether there will be a loss or gain? The paradox illustrates one of the fundamental properties of the Fibonacci sequence. If any number in the sequence is squared, it equals the product of the two numbers on either side of it, plus or minus 1. Expressed algebraically,

$$t_n^2 = (t_{n-1} \cdot t_{n+1}) \pm 1$$

The left side of the above equation clearly gives the area of the square, and the right side clearly gives the area of the rectangle. The plus and minus signs alternate throughout the sequence. Every Fibonacci number located at an odd position in the series (for example, 2, 5, or 13 in the Fibonacci series above) has a square that is 1 greater than the product of the two adjacent numbers in the even positions on either side. Conversely, every number located in an even position (for example, 3, 8, or 21 in the Fibonacci series above) has a square that is 1 less than the product of its two adjacent numbers in the odd positions on either side. Once you know this, it is easy to predict whether the rectangle of a particular square pattern will gain or lose a unit of area.

The Fibonacci sequence starts with 1, 1, but a "generalized Fibonacci sequence" can start with any pair of numbers. You can explore variants of the paradox based on other Fibonacci sequences. For example, the sequence 2, 4, 6, 10, 16, 26, . . . gives losses and gains of 4 square units. The sequence 3, 4, 7, 11, 18, . . . gives losses and gains of 5 square units.

Let a, b, c stand for any three consecutive terms in a generalized Fibonacci sequence, and x for the loss or gain. Two formulas hold:

$$a + b = c$$
$$b^2 = ac \pm x$$

We can substitute for x whatever loss or gain we desire, and for b whatever length we wish for the side of the square. Solving the two simultaneous equations then provides values for a and c, though they may not be rational numbers.

Can the square be cut in such a way that when the four pieces are rearranged, the rectangle will have precisely the *same* area as the square?

To answer this, let $x = 0$ in the second of the above two equations, and solve for b in terms of a. The only positive solution is

$$b = \frac{(1 + \sqrt{5})\,a}{2}$$

The expression $(1 + \sqrt{5})/2$ is the famous golden ratio, or phi, written ϕ. This is an irrational number equal to 1.618033. . . . In other words, the only Fibonacci sequence in which the square of a term exactly equals the product of its two adjacent numbers is

$$1, \phi, \phi^2, \phi^3, \phi^4, \ldots$$

With some manipulation of radicals we could prove that the above sequence is a true Fibonacci sequence by showing it is equivalent to

$$1, \phi, \phi + 1, 2\phi + 1, 3\phi + 2, \ldots$$

Only by cutting the square with lengths that are consecutive numbers in the above sequence can we produce a variant of the paradox for which the areas of the square and the rectangle are identical. For more on the golden ratio, and its relation to the square–rectangle paradox, see the chapter on ϕ in my *Second Scientific American Book of Mathematical Puzzles and Diversions*.

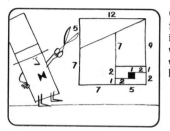

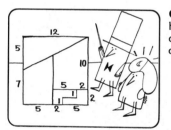

A few months later, Mr. Randi returned with a rug 12 decimeters by 12 decimeters.

Randi: Omar, old pal, my electric heater overturned and burned this beautiful carpet. By cutting and sewing, it will be easy to get rid of the hole.

Omar was doubtful, but he followed Mr. Randi's instructions. After the pieces were sewn together, the rug was still 12-by-12 but the hole had vanished!

Omar: Please, Mr. Randi, how did you do it? Where did that square decimeter come from to fill the hole?

How can two identical squares have different areas? In Randi's second rug paradox, the loss of area appears as an actual hole. Unlike the previous paradox, there is an accurate fitting along the sloping line in both patterns. What happens to that missing square unit?

To find the answer, make two copies of the square without the hole. The larger the patterns the better. One square should then be cut accurately, the pieces rearranged to make the hole, and this second pattern placed on top of the first. If the top and sides are flush, you discover that the second pattern is not a true square. It is a rectangle that is higher than the square by a length of 1/12 decimeter. This 12-by-1/12 strip along the bottom has the same area as the hole.

That explains where the missing square unit goes. But why does the square grow in height? The secret is that the vertex on the hypotenuse of the triangular piece is not on a lattice point. Knowing this you can construct variants of the square in which the loss or gain of area is more than 1 square unit.

The paradox is known as a Curry square after its inventor, Paul Curry, an amateur New York magician. It has numerous variations, including triangular forms. If you want to know more about Curry squares and triangles, see Chapter 8 of my *Mathematics, Magic and Mystery* and Chapter 11 of my *New Mathematical Diversions from Scientific American*.

The Vanishing Leprechaun

The funniest versions of these paradoxes are those in which a drawing of a person is caused to disappear. Consider, for example, The Vanishing Leprechaun Puzzle, drawn by Pat Patterson of Toronto, copyrighted and sold by the Elliott Company of Toronto. The puzzle is reproduced below. To avoid damaging the book, photocopy it, then cut the border and along the dotted lines to make three rectangles. Switch the two top rectangles, and one of the fifteen leprechauns disappears without a trace! Which one vanished? Where did he go? When he comes back, where has he been?

If you would like to obtain a deluxe version of this paradox, printed in color on cardboard 19 inches long, write for the current price to W. A. Elliott Company, 212 Adelaide Street West, Toronto, Ontario, Canada M5H 1W7.

Vanishing-person paradoxes have been used for more than a century as advertising premiums. In the 1880s, the American puzzle inventor Sam Loyd issued a circular version in which a Chinese warrior seems to vanish when a disk is rotated. Many other versions, both straight and circular, have been printed since.

The best way to explain the paradox is to rule ten lines on a card like this:

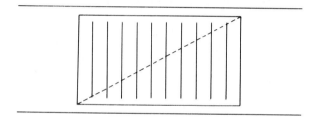

Cut the card along the dotted line, and then slide the lower part down and left.

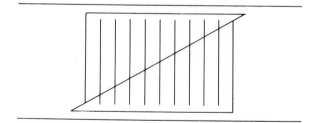

Count the lines. There are only nine! It is meaningless to ask which of the original ten lines vanished. What has happened is that the ten lines have been broken into 18 parts, and those parts rearranged to make a new set of nine lines. Each of the nine lines is, of course, longer by 1/9 than each of the ten lines. When we slide the lower part back up again, a tenth line appears, and now the lines are shorter by 1/10 than they were before.

Exactly the same thing happens with the leprechauns. When there are 15 leprechauns, each is shorter by 1/15 than when there are 14. We cannot pick out a leprechaun that vanishes when we change the pieces, because the set of 14 is a set of *completely different* leprechauns. Each is longer by 1/14 than before. For a lengthy discussion of this paradox, and others related to it, see Chapter 7 of my *Mathematics, Magic and Mystery*.

The principle behind this paradox is also the basis of an old counterfeiting method. It is possible to cut 9 bills into 18 parts (following the pattern of the lines) and to rearrange them to make 10 bills. However, the new bills are easy to detect because their numbers will not match. The two numbers on all U.S. bills are placed on opposite ends, one high and one low, precisely to foil this counterfeiting scheme. In 1968, a man in London was sentenced to eight years in prison for using the scheme on British 5-pound notes.

The Great Bank Swindle

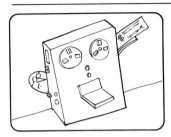

Believe it or not, these paradoxes have something in common with the method used by an unscrupulous computer programmer to steal from a large bank.

Thief: Man, am I a genius! I'll be ripping off $500 a month, and it's so easy! I've just told our computer to round *down* every customer's account instead of rounding up or down to the nearest penny.

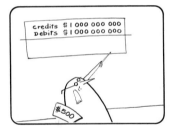

Thief: Each customer will lose about half a penny every month. That's too small for anyone to notice. But the bank has a hundred thousand customers, so the total loss is $500. Each month the computer will deposit it to my secret account, and the books will always balance!

Vanishing-area paradoxes operate by stealing tiny bits of area from many spots. Randi's first rug, after its pieces are rearranged, has an imperceptible overlap along the rectangle's main diagonal. Randi's second rug, after it is cut and reformed, shrinks in height by a trifle amount. After a leprechaun disappears, each leprechaun is slightly taller than the previous figures. After $500 appears in the thief's account, some of the customers' accounts were credited a penny less interest than they should have been.

The Amazing Inside-Out Doughnut

Topology is called rubber-sheet geometry because it studies properties that do not change when a figure is stretched or distorted.

A torus is a fascinating surface shaped like a doughnut. Would you believe that a torus made of thin rubber, with a big hole in it, can be turned inside out through the hole? It *can* be done but it's very difficult.

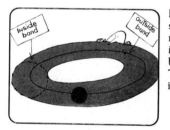

Before the torus is reversed, suppose we glue a small rubber band around its *inside* and a larger rubber band around its *outside*. The two bands are *not* interlocked.

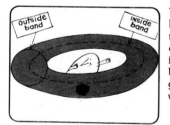

This is how the torus should look after it has been reversed. But now the two elastic bands are linked! It's impossible to link two bands without cutting and gluing, so *something* is wrong. But what?

It is true that a torus can be turned inside out through a hole, but this will *not* interlink the two rubber bands. The reason is that when the torus is reversed, the bands change places! After the reversal, the small band has been stretched to the larger one and the large band has shrunk to the smaller, so that the bands are as unlinked as before. The key to the paradox lies in the fact that the artist deliberately drew the second picture to show what we would expect to see, rather than what actually happens.

A rubber model of a torus, such as an inner tube, is very difficult to turn inside out through a hole because the rubber has to be stretched so radically. But it is easily done with a cloth model. Fold a square piece of cloth in half and sew opposite edges together to make a tube. Now fold the cloth the other way and sew the opposite ends of the tube together to make a torus. It will be square-shaped when flattened. For ease in reversing, the "hole" is a slot cut horizontally in the outer layer of cloth.

Turning this cloth torus inside out through the slotted hole is easy. After the reversal, the torus has the same shape as before, except now the slot is vertical, and the grain of the cloth has also been turned 90 degrees. In other words, lines that previously circled the torus one way now circle it the other way. To see how this turning of the "grain" explains the switch of the two elastic bands, you can use two felt-tip pens to ink a ring of one color around the torus one way, and a ring of another color around the torus the other way. After turning the torus inside out, you will see that the rings have exchanged positions.

It is not easy to visualize exactly how the torus is distorted during the reversal process. A series of drawings, showing all stages of the reversal, can be found in "Topology" by Albert Tucker and Herbert Bailey in *Scientific American*, January 1950, and on p. 179 of *Mathematics*, Life Science Library.

There are many other torus paradoxes. For example, if a torus without a hole is linked to a torus with a hole, can the one with the hole "swallow" the other torus so that it is completely inside? The answer is yes, and you will see how it is done if you consult my *Scientific American* column of March 1977. More paradoxes involving toruses are in my columns of December 1972 (on knotted toruses) and December 1979.

The Bewildering Braid

Wendy is shopping for a leather bracelet.

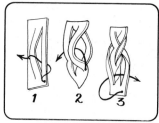

It looks impossible, but Luke braided the bracelet in 30 seconds, and without cutting a single thong! Here's how he started.

Inside Luke's shop she sees two bracelets, each made of 3 leather thongs. One is braided, the other is not.
Wendy: How much is that braided one?

Luke: Five dollars, ma'am. But you're too late. I've just sold it.
Wendy: Oh dear! Do you have another?

Luke: Yes, I have this one.
Wendy: But it's not braided.
Luke: I'll be happy to braid it for you, ma'am.

The astonishing thing about the bracelet is that the six-crossing braid can be formed even though the ends of the strands are permanently attached to each other. In other words, the braided bracelet is topologically equivalent to the unbraided one. The picture below shows one procedure for forming the braid. By repeating it, with longer strands, one can extend the braid to any multiple of six crossings. To make an actual bracelet or a braided belt from a piece of stiff leather, first soak it in warm water to make the leather pliable.

Braids of this sort can be made with more than three strands. More information can be found in J. A. H. Shepperd's "Braids Which Can Be Plaited with Their Threads Tied Together at Each End" (*Proceedings of the Royal Society, A,* vol. 265 (1962), pp. 229–244). See also the chapter, "Group Theory and Braids," in my *New Mathematical Diversions from Scientific American.*

Most people will see this bracelet as just another topological curiosity, but it is much more than that. Emil Artin, a famous German mathematician who settled in the United States, developed a theory of braiding in which he applied group theory. Thus, the "elements" of the group are "weaving patterns," the "operation" consists of following one pattern with another, and the "inverse" of a weaving pattern is its mirror image. Braids provide an excellent jumping-off point for discovering groups and transformations. (A good introduction to braid theory is Artin's article "The Theory of Braids" in *The Mathematics Teacher,* May 1959.)

The Inescapable Point

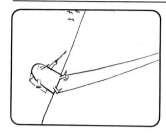

Pat hiked up a narrow trail that leads to a mountain top. He started at 7:00 that morning and arrived at the summit at 7:00 that night.

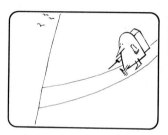

After a night of meditation on the summit, Pat started back down the trail at 7:00 in the morning.

At 7:00 that evening, when he reached the bottom, Pat happened to meet Ms. Klein, his topology teacher.
Ms. Klein: Hello, Pat. Did you know that coming down today you passed a certain spot at *precisely* the same time as you did yesterday on your way up?
Pat: You must be putting me on. It can't be! I walked at different speeds. I even stopped to eat and rest.

But Ms. Klein was right.
Ms. Klein: When you started *up* the mountain, suppose you had a double who started *down* at the same time. No matter what your double's schedule, somewhere along the trail, the two of you are sure to meet.

Ms. Klein: We can't say where you'll meet, but we know it will be *somewhere.* You and your double will be there at the same time. So there has to be a point on the trail that you pass at the same instant on your up and down trips.

If for each point on the path we pair Pat's arrival time going up with his arrival time going down we get a correspondence between times. At least one of these times must be matched up with itself. Thus the story about Pat illustrates a very simple example of what topologists call a *fixed-point theorem*. The proof is an "existence proof." It does no more than establish the existence of at least one fixed point. It provides no help in locating the fixed point. Fixed-point theorems are extremely important in the application of topology to many branches of mathematics and science.

A famous fixed-point theorem can be demonstrated with a shallow box and a piece of paper that exactly covers the bottom of the box. Imagine that every point on the paper is paired with a point directly beneath it on the bottom of the box. Pick up the paper, crumple it into a ball, and drop it back in the box. Topologists have proved that no matter how the paper is crumpled, or where it falls in the box, there must be at least one point on the paper that is directly above its corresponding point on the box! See "A Fixed Point Theorem" in *What is Mathematics?* by Richard Courant and Herbert Robbins.

This theorem, first proved by the Dutch mathematician L. E. J. Brouwer in 1912, has many curious applications. For example, it establishes that at every instant there is at least one spot on the earth where no wind is blowing. One can also prove that on the earth there are always at least two antipodal points (points joined by a straight line through the earth's center) that have exactly the same temperature and barometric pressure. A similar theorem can also be used to prove that if a sphere is completely covered with hair it is impossible to comb all the hairs flat. (One *can* comb flat all the hairs on a doughnut.) For a good introduction to such theorems see "Fixed-Point Theorems," by Marvin Shinbrot, in *Scientific American,* January 1966.

Impossible Objects

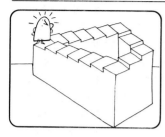

If Pat was surprised by that point, he'd be even more surprised by this stairway. He can walk around it forever, always climbing up, but always coming back to where he started!

Are there two or three prongs on the knight's weapon?

Could you build a model of this crazy crate?

The stairway, the weapon, and the crate are called "impossible objects" or "undecidable figures." The impossible stairway was invented by the British geneticist Lionel S. Penrose and his mathematician son Roger Penrose, who first published it in 1958. It is often called the Penrose staircase. The Dutch artist M. C. Escher was fascinated by it. He made effective use of it in one of his lithographs, *Ascending and Descending*.

The impossible figure of two or three prongs is of unknown origin. It began circulating among engineers and others about 1964. *Mad* Magazine, March 1965, had a cover that showed Alfred E. Neuman balancing one of these objects on his index finger.

The crazy crate, also of unknown origin, appears in another Escher picture, *Belvedere*. All three objects show how easily we can be tricked into thinking that a geometrical diagram represents a genuine structure when the structure is logically contradictory and therefore cannot exist. The objects are visual analogs of such undecidable sentences as "This sentence is false" discussed in Chapter 1.

For more examples of undecidable figures see the chapter on optical illusions in my *Mathematical Circus* and the books of the Japanese graphic artist Mitsumasa Anno, especially *Anno's Alphabet* and *Anno's Unique World*.

A Pathological Curve

The snowflake curve is another paradoxical figure, but it is not impossible. We start constructing it with the shape of this Christmas tree—an equilateral triangle.

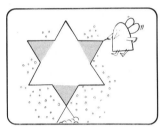

By drawing a shaded equilateral triangle on the central third of each side of the light one, this little angel produces a six-pointed star.

He repeats the construction by drawing still smaller triangles on the sides of the star. The curve is getting longer and starting to look like a snowflake.

The next repetition makes the curve still longer and prettier.

Continuing in this way, the curve becomes as long as one likes. It can be drawn on a postage stamp, yet made as long as the distance from earth to the farthest star!

The snowflake curve is one of the prettiest of an infinite class of curves called *pathological* because of their paradoxical properties. If the construction of the snowflake is continued ad infinitum, its length at the limit is infinite, yet it encloses a finite area! Put another way, the successive lengths of the curve, at each step, form a divergent series, but the successive areas enclosed by the curve form a series that converges on an area 8/5 of the original triangle. Moreover, it is impossible to define a tangent for any point on the limit curve.

The snowflake curve is a good way to consolidate the concept of limit. Can you show that if the starting triangle has an area of 1, the curve's limiting area is 8/5?

Here are some related constructions:

1. Construct the anti-snowflake by drawing the triangles inward instead of outward, and erasing the base lines. The first step produces three diamonds meeting at a point like three propeller blades. Is this also a curve that is infinite in length at the limit? Does it also enclose a finite area?

2. What happens when you use other regular polygons as the basis of construction?

3. Investigate the effect of constructing more than one polygon on each side.

4. Are there 3-dimensional analogs of the snowflake and its cousins? For example, if tetrahedrons are constructed on the faces of tetrahedrons, will the limit solid have an infinite surface area? Will it bound a finite volume?

For a paradoxical curve discovered by William Gosper, called the "flowsnake," see my column on pathological curves in *Scientific American,* December 1976. Another remarkable curve, discovered by Benoit Mandelbrot, was depicted on the cover of *Scientific American,* April 1978, and discussed in my column for the same issue. See Mandelbrot's book *The Fractal Geometry of Nature* for more on pathological curves related to the snowflake.

The Unknown Universe

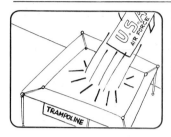

If a spaceship blasted off and kept going in a straight line, would it get farther and farther from the earth? Maybe not, suggested Einstein. It might come *back* to earth!

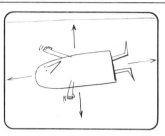

The Flatlander lives on a surface of two dimensions. If his universe is an infinite plane, he can travel on it forever in any direction.

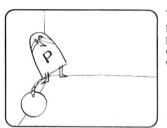

To understand Einstein's paradox, first consider this poor Pointlander. He lives on a single point. His universe has *no* dimensions.

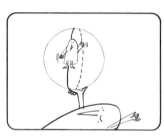

But if his surface is closed like that of a sphere, it too becomes finite and unbounded. He too will return to where he started if he travels along a straight line path in any direction.

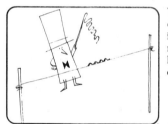

The Linelander who lives on a line of one dimension is like the worm on this rope. If the rope is infinite, he can travel forever in either direction.

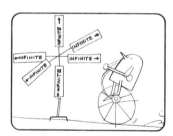

You and I are Solidlanders who live in a 3-dimensional space. Perhaps it is infinite in all directions.

However, if the rope is closed like a circle, it becomes a line that is unbounded, yet finite in length. If the worm crawls either way, it will return to its starting point.

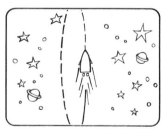

Or, as Einstein thought, it may curve through a higher space to form another finite but unbounded universe. A spaceship speeding through such a universe, along the straightest possible path, would eventually return home.

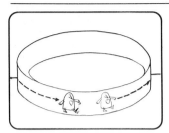

When a Flatlander circles a sphere, it is like walking around an untwisted band. If he has a heart on one side, it stays on the same side.

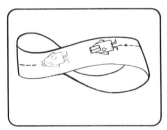

But if a Flatlander goes around a Moebius band, something strange happens. The twist flips him over so that he returns with his heart on the *other* side!

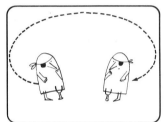

If our space is closed, it could be twisted the way a Moebius strip twists. An astronaut who made a round trip through such a space would come back reversed.

Astronomers do not yet know whether the space of our universe is closed, as Einstein suggested, or open. It all depends on how much matter is in the universe. According to the general theory of relativity, the presence of matter in space causes space to "curve," and the amount of curvature increases as the proportion of matter increases. Most cosmologists today believe there is not enough matter to produce a curvature great enough to close space. However, the issue is still open because the nature and density of matter in the universe are not yet known. The universe may contain enough "hidden matter" to close itself. (At the moment, neutrinos are suspected of having positive rest mass rather than having a rest mass of zero as was formerly believed).

There is no evidence that the space of our universe twists like a Moebius strip, but cosmologists like to invent different models of the cosmos, and in some of these models, space has a twist. In understanding how a Flatlander would get mirror-reversed by going once around a Moebius strip, it is important to realize that the strip has zero thickness. A paper model of a Moebius surface is really a solid, because the paper has thickness. We must assume that the true Moebius surface has no thickness.

A plane figure drawn on a Moebius strip is like a figure drawn with ink that soaks through the paper so that it is on both "sides," not a figure that slides along one side only. It is "embedded" in the surface. When it makes one trip around the strip, it arrives at its starting point in reversed form. Of course, a second trip around the strip brings it back to unreflected form. In the same way, if an astronaut made a round trip through a twisted cosmos, he would return reversed, but a second trip would straighten him out again.

If you are intrigued by the paradoxical properties of the Moebius strip, you may wish to investigate two other surfaces equally paradoxical: the Klein bottle and the projective plane. Both are one-sided, but unlike the Moebius surface they have no edges. Both are closed like the surface of a sphere. The Klein bottle is closely related to the Moebius strip because it can be cut in half to make two Moebius strips that are mirror images of one another. A Flatlander embedded in either a Klein bottle or a projective plane can mirror-reflect himself by making a trip around the surface (see Chapter 2 of my *Sixth Book of Mathematical Games from Scientific American*). The classic book about life in two dimensions is Edwin A. Abbott's *Flatland*. A sequel, *Sphereland,* was written by Dionys Burger.

You might also enjoy a story by H. G. Wells called "The Plattner Story" (in his collection *28 Science Fiction Stories*). This science fiction tale is about a man who gets himself reversed in outer space and returns with his heart on the wrong side.

Antimatter

A reversed astronaut would feel normal, but the world around him would seem mirror-reflected. Printing would go the wrong way. Cars would be on the wrong side of the road.

Many physicists believe that reflected matter would become antimatter, which is annihilated when it contacts ordinary matter. If so, our reversed astronaut could never get back to the earth. As soon as his reversed spaceship touched our atmosphere, it would explode!

Does our universe contain galaxies of antimatter? Are there vast universes of anti-matter outside our own? Cosmologists today can only guess at the answers.

Every fundamental particle has an antiparticle that is the same as the particle except that its electric charge (if it has one) is reversed, as well as other properties. Many physicists believe that an antiparticle is simply a particle whose internal structure is mirror-reflected. Matter made of antiparticles is called antimatter.

When a particle meets an antiparticle, there is mutual annihilation. Our galaxy is made entirely of matter, so whenever an antiparticle is created, either in the laboratory or in the interior of stars, it lasts only for a microsecond before it is destroyed by meeting its opposite particle.

Most cosmologists believe that the universe consists only of matter, but a few have argued for the possibility that it may contain galaxies of antimatter. Light from such galaxies would be indistinguishable from light from galaxies made of matter, so it is difficult to know. Some cosmologists have speculated that immediately after the Big Bang, which presumably started the universe evolving, matter and antimatter may have separated to form two universes: a "cosmon" and an "anticosmon," which repelled each other and separated at great speed.

The notion that the universe is divided into these two parts, each a kind of mirror image of the other, has played a role in many science fiction stories and in Vladimir Nabokov's romantic novel *Ada.* You can find more about antimatter and related topics by reading Chen Ning Yang, *Elementary Particles;* Hannes Alfvén, *Worlds—Antiworlds;* and my *Ambidextrous Universe.*

Probability

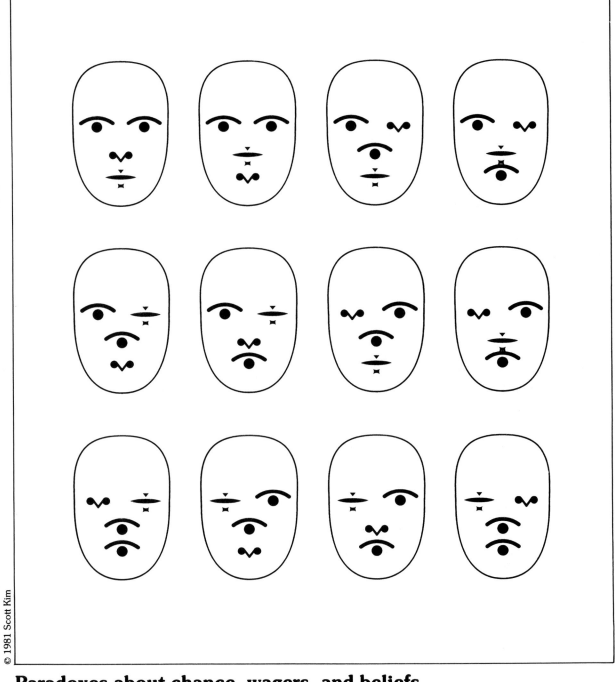

Paradoxes about chance, wagers, and beliefs

Probability theory has become so essential in every branch of science, not only in the physical sciences but also in the biological and social sciences, that it is safe to predict that in the years to come it will be emphasized more and more in the teaching of mathematics in elementary grades. Bishop Joseph Butler and others before him (Cicero, to mention one) have said that probability is the very guide of life. From morning until night we live by making thousands of unconscious little bets about probable outcomes. If quantum mechanics is the final word in physics, pure chance underlies all of nature's fundamental laws.

More than most branches of mathematics, probability swarms with results that are strongly counterintuitive, with problems for which the correct solution seems utterly contrary to common sense. If you walked up to an elevator door, you might expect the chances to be fifty–fifty that the first time the elevator stops it will be on its way up. Paradoxically, this is generally false. In a family with four children you might expect that the most likely situation would be two children of each sex, but this is also false.

The simple ideas of probability introduced here will help you understand why the bets that appear favorable in chuck-a-luck are actually unfavorable. More generally, these ideas are useful in understanding why amazing coincidences really are not so amazing, but we leave that to the next chapter.

The paradoxes in this chapter have been selected because they are easy to understand and because many of them can be modeled with such readily accessible materials as coins and playing cards. Wherever possible, a paradox has been explained by listing all the equally possible cases even though the problem can be solved in shorter ways using probability theory. By solving them the longer way, you gain an insight into the problem's structure that is not obtainable otherwise.

Although ultimately there may be only one kind of probability, it is customary these days to distinguish at least three main types:

1. Classical or a priori probability. Here we assume that each outcome is equally likely. If an event is found to have n equally likely outcomes, and you want to know the probability that a certain subset k of those outcomes will take place, the answer is the fraction k/n. For example, a rolling die, if the die is fairly made, has six faces each equally likely to show on top. What is the probability you will roll an even number? Of the six equiprobable cases (1, 2, 3, 4, 5, 6) three are even (2,4,6), therefore the probability of rolling an even number is $3/6 = 1/2$. Put another way, the odds are even. It is a fair bet.

2. Frequency or statistical probability. This concerns events that do not seem, a priori, to be equally probable. The best we can do is repeat or observe the event many times and note the frequency with which certain outcomes occur. An example would be a die loaded in a manner that cannot be easily determined by inspection. So you roll it hundreds of times. By keeping records you conclude that the probability of rolling, say, a 6 is about $7/10$ instead of the familiar $1/6$ for a fair die.

3. Inductive probability. This is the degree of credibility a scientist assigns to a law or theory. Insufficient knowledge of nature precludes a classical solution, and experiments or observations are too infrequent and vague to permit accurate frequency estimates. For example, a scientist considers all the relevant evidence based on the scientific knowledge of his time and concludes it is more likely than not that black holes exist in the universe. Such probability estimates, necessarily imprecise, constantly change as new evidence is found that bears on the hypothesis.

Our last two paradoxes touch on inductive probability, as do the last two paradoxes of the next chapter. If you read more about such paradoxes, you will find yourself plunging into some of the deepest waters of modern probability theory and the philosophy of science.

The Gambler's Fallacy

Mr. and Mrs. Jones have five children, all girls.
Mrs. Jones: I do hope our next child isn't another girl.
Mr. Jones: My dear, after five females, it's bound to be a boy.
Is Mr. Jones right?

The probability of Mr. and Mrs. Jones having a sixth girl is the same as that of her first child having been a girl. The probability of the next roulette number being red is the same as that of the previous number having been red. The probability of a two on the next throw of the die is still one-sixth.

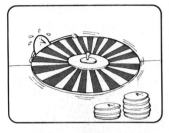

Many gamblers think they can win at roulette by waiting until there is a long run of red numbers, then betting on black. Will such a system work?

To make this clearer, suppose Mr. Jones tosses five heads in a row with a fair coin. The chances of tossing heads again is exactly the same as before: fifty-fifty. The coin has no memory of what it did on previous flips.

Edgar Allan Poe argued that if you roll five twos in a row, your chances of getting a two on the next roll are less than one-sixth. Was he right?

If you answered yes to any of these questions, you have fallen into a trap called the "gambler's fallacy." In every case the next event is completely independent of all previous events.

If the outcome of event A influences event B, then B is said to be "dependent" on A. For example, the probability of your wearing a raincoat tomorrow clearly depends on the probability of rain tomorrow, or (more directly) on how you estimate that probability. Events that in ordinary language are said "to have nothing to do with each other" are called "independent" events. The probability of your wearing a raincoat tomorrow is independent of the probability that the President of the United States has eggs tomorrow for breakfast.

Most people find it difficult to believe that the probability of an independent event is not somehow influenced by its proximity to other independent events of the same sort. During World War I, for example, soldiers on the front looked for fresh shell holes to hide in. They were convinced that old shell holes were risky because they believed it would be high time for new shells to land in the same spot. Because it seems unlikely that two shells would fall on the same spot, one right after the other, they reasoned that a fresh shell hole would remain safe for quite some time.

A story was told many years ago about a man who traveled a lot in airplanes. Fearful that a passenger might someday take aboard a concealed bomb, he always carried in his briefcase an unloaded bomb of his own. He knew it was unlikely that a plane would have *one* passenger with a bomb, so he reasoned it would be very much more unlikely that a plane would have *two* passengers with a bomb. Of course carrying his own bomb would have no effect on the probability that another passenger had a bomb, any more than the flip of one penny can be influenced by the flip of another penny.

The most popular of all roulette systems, known as the d'Alembert system, is based squarely on the "gambler's fallacy" of not recognizing the independence of independent events. The player bets on red or black (or makes any other even-money bet), following it with a larger bet after each loss and a smaller bet after each win. The assumption is that if the little ivory ball allows him to win, it will somehow "remember" that and be less likely to let him win the next time. And if the ball causes him to lose, it will feel sorry for him and be more likely to help him on the next spin of the wheel.

The fact that each spin of a fair roulette wheel is independent of all previous spins provides a very simple proof that no roulette system can give a player an advantage over the house. The word *odds* can be used with two meanings. The odds that a fair coin will fall heads up are even or 1 to 1 (or 5 to 5). But a bookie, mindful of his need to turn a profit, might pay off $4 for your bet of $5 on heads. He says, "The odds on heads are 4 to 5." He would be offering you less than correct odds. In his *Complete Guide to Gambling*, John Scarne puts it this way:

When you make a bet at less than the correct odds, which you always do in any organized gambling operation, you are paying the operator a percentage charge for the privilege of making a bet. Your chance of winning has what mathematicians call a "minus expectation." When you use a system, you make a series of bets, each of which has a minus expectation. There is no way of adding minuses to get a plus. . . .

Edgar Allan Poe's howler about dice occurs in the postscript to his detective story, "The Mystery of Marie Roget." A die, like a penny, a roulette wheel, or any other randomizing device, produces a series of independent events that are not influenced in any way by the device's previous behavior.

If you are inclined to believe in some form of the gambler's fallacy, test it by simulating an actual game in which you play a system based on the fallacy. For example, toss a penny repeatedly, betting a poker chip, at even odds, only after runs of three showings of the same side. Always bet on the coin's opposite side. In other words, after three heads, bet tails, and after three tails bet heads. At the end of, say, 50 such bets it is unlikely you will have *exactly* the same number of chips you began with, but it should be close. The probability of being ahead or behind is, of course, equal.

Four Kittens

It's easy to go wrong calculating probabilities. Here are two cats that have been hanging out together.

Mrs. Katz: Or maybe just one is a girl.

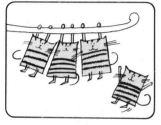

Mr. Katz: Love, how many kittens are in our new litter?
Mrs. Katz: Can't you count? Four, you big dope.
Mr. Katz: How many boys?
Mrs. Katz: It's hard to tell. I don't know yet.

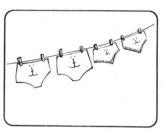

Mr. Katz: It's not so hard to figure out. There's a fifty-fifty chance each kitten is a boy or girl. So it's obvious that the most likely outcome is two boys and two girls. Have you named them yet?

Mr. Katz: It's not very likely that all four are boys.

Has Mr. Katz reasoned correctly? Let's check out his theory. Using "B" for a boy and "G" for a girl, list all 16 of the equally possible cases.

Mrs. Katz: And it's not likely they're all girls.

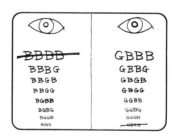

Only 2 of the 16 cases show all kittens of the same sex. So the probability of this happening is 2/16 or 1/8. Mr. Katz was right in thinking that this outcome had a low probability.

Mr. Katz: Maybe just one is a boy.

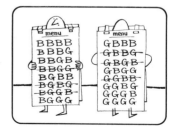

Now let's check the 2–2 split that Mr. Katz thought the most likely. It happens six times. So the probability is 6/16 or 3/8. This is surely higher than 1/8. Mr. Katz may be right.

But we have one more possibility to consider: a 3–1 split. Because this occurs in eight cases, the probability is 8/16 or 1/2. This is even higher than the 2–2 split. Could we have made a mistake?

All of same sex ⅛
A 3–1 split 4/8
A 2–2 split 3/8
8/8 = 1

If our probabilities are correct, they should add to 1. They do. This tells us it is certain that one of the three splits will occur. Mr. Katz's guess was wrong. The most likely split is not 2–2, but 3–1.

That four children in a family are more likely to consist of three of one sex and one of another, than to consist of two boys and two girls, is surprising to most people. It is easily tested by repeatedly tossing four pennies. Keep a record of each toss. After a hundred tosses, approximately 50 should show a 3–1 split compared to about 33 that show a 2–2 split.

You may be curious about the probabilities of the different sex distributions in families of five and six children. These can be found by listing all combinations, but this is tedious. You might find it easier to use shorter methods found in books on probability.

A similar problem, with an equally counter-intuitive answer, concerns the most likely way that the four suits are distributed in a bridge hand. The least likely, of course, is to hold 13 cards of one suit. (The odds against this happening to you are 158,753,389,899 to 1.) But what distribution of suits is *most* likely to occur?

Even seasoned bridge players often guess the answer to be 4, 3, 3, 3. This is incorrect. The most probable hand is a 4, 4, 3, 2 distribution. You can expect to get a hand of this sort about once every five deals compared to once every nine or ten deals for a 4, 3, 3, 3 hand. Even a 5, 3, 3, 2 hand is likely to occur about once in every six deals. For a chart giving the probabilities of all possible suit distributions, see Oswald Jacoby, *How to Figure the Odds.*

Every now and then you see a newspaper story about someone getting a perfect bridge hand. The odds against this are so astronomical that the story is almost certainly a hoax, or someone at the table was a practical joker who secretly arranged the cards, or perhaps a new deck was opened and someone accidently gave it two perfect riffle shuffles. A perfect shuffle is one that divides the deck exactly in half, then interlaces alternate cards. New decks come with the four suits separated. Two perfect shuffles, followed by any kind of cut, will set a deck for dealing *four* perfect bridge hands.

Three-Card Swindle

In many gambling games, trusting your intuition about probability can be disastrous. A simple betting game with three cards and a hat proves it.

The mirror reflection makes it easy to see how the cards are made. The first card has a spade on both sides. The last card has a diamond on both sides. The middle card has a spade on one side and a diamond on the other.

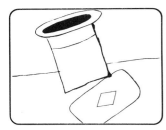

The banker shakes the cards in a hat and lets you pick one and put it on the table. He then bets even money that the underside suit is the same as the top. Suppose the top of your chosen card is a diamond.

To con you into thinking it a fair game, the banker tells you that your card cannot be the spade–spade card. Therefore it is either the spade–diamond card, or the diamond–diamond card. One has a diamond on the bottom, the other a spade, so you and he have equal chances of winning.

If the game is fair, how is it that the banker so quickly rakes in your money? It's because his argument is phony. The actual odds are two to one in his favor!

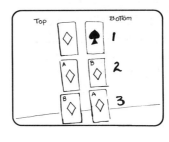

The catch is that there are *three,* not two, equally likely cases. The drawn card can be spade–diamond, or diamond–diamond with side A on top, or diamond–diamond with side B on top. The bottom matches the top in two cases. Therefore, in the long run the banker wins two out of three games.

This card-betting game was designed by Warren Weaver, the distinguished mathematician who is one of the co-founders of information theory. He introduced it in the article "Probability" in *Scientific American*, October 1950.

One way of explaining the game's true odds is given above. Here is another. Two cards each have matching colors. If you take a card at random from the hat, the probability is 2/3, or 2 out of 3, that you will get one of those two cards. Therefore, the probability is 2/3 that the underside of the selected card will match its top.

The card game is a variation of what is known as Bertrand's box paradox, after J. Bertrand, a French mathematician who presented it in a book on probability in 1889. Bertrand imagined three boxes. One contains two gold coins, one contains two silver coins, and one has a gold coin and a silver coin. A box is picked at random. Clearly the probability is 2/3 that this box contains matching coins.

Suppose, however, that we take one coin from the chosen box and observe that it is gold. This tells us that the box cannot be the silver–silver box. Therefore it must be the gold–gold or the gold–silver box. Since each of these two boxes is equally likely to have been chosen, it seems as if the probability we have taken a box of matching coins has gone down to 1/2. The same argument would apply if our sample coin had been silver.

How can looking at one coin in the box alter the probability of that box holding matching coins? Clearly it cannot.

Here's a related paradox. What is the probability that if you toss three coins, all of them fall alike? At least two of the three *must* be alike. The third coin will either match that pair or be different. Since the chances are fifty–fifty it will fall either way, the chances are even that it will match. Therefore the probability all three coins are alike seems to be 1/2.

We can show this reasoning to be wrong by listing the eight possible cases:

HHH THH
HHT THT
HTH TTH
HTT TTT

Observe that only two of the cases show all three coins the same. The correct probability, therefore is 2/8 = 1/4.

Another bewildering little paradox, again arising from a failure to consider all possible cases, involves a boy with one marble and a girl with two marbles. They roll their marbles toward a stake in the ground. The person whose marble is closest to the stake wins. Assume that the boy and the girl are equally skillful and that measurements are accurate enough to eliminate ties. What is the girl's probability of winning?

Argument 1: The girl has two marbles to roll, against the boy's one, therefore her probability of winning is 2/3.

Argument 2: Call the girl's marbles A and B, and the boy's marble C. There are four possible results:

1. Both A and B are closer to the stake than C.
2. Only A is nearer than C.
3. Only B is nearer than C.
4. C is nearer than A and B.

In three of the four cases the girl wins, therefore her winning chances are 3/4.

Which argument is correct? To settle the matter, we make an exhaustive list of not four but six possible cases. The equally possible orderings of marbles, listing the nearest marble first, are:

ABC
ACB
BAC
BCA
CAB
CBA

In four of the six cases the girl wins. This confirms the original argument that her chances are 4/6 = 2/3.

The Elevator Paradox

People who ride elevators are often puzzled by another strange probability paradox. We'll assume that elevators in this building move independently and average the same waiting time on each floor.

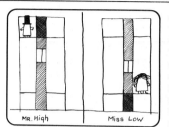

A simple diagram clears up the mystery. For Mr. High, only elevators in the dark region of their shafts are going down. This region is small compared to the light region, so the probability is much higher that an elevator is below his floor and coming up. The same argument works in reverse for Miss Low.

Mr. High has an office on a floor near the top. He's very annoyed.
Mr. High: Confound it! The first elevator to stop here is going up. It happens all the time.

Mr. High: Maybe they're making elevators in the basement and taking them off the roof in helicopters.

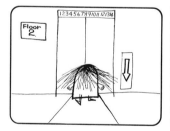

Miss Low works on a floor near the bottom. Every day she eats lunch in a restaurant on the top floor. She, too, is annoyed.
Miss Low: I can't understand it. Whenever I want an elevator, the first one to arrive is usually going down.

Miss Low: They must be bringing elevators to the roof, then sending them down to be stored in the basement.

The elevator paradox first appeared in the book *Puzzle-Math* by the physicist George Gamow and his friend Marvin Stern. In explaining the paradox with one elevator, Gamow and Stern made a small mistake. They stated that the probabilities "of course remain the same" if there are two or more elevators.

Donald Knuth, a Stanford University computer scientist, was the first to realize that this is not true. Writing on "The Gamow–Stern Elevator Problem" in *The Journal of Recreational Mathematics* (July 1969), Knuth showed that as the number of elevators increases, the probability that the first elevator to stop on any floor is going up approaches 1/2, and the probability it is going down also approaches 1/2.

This situation, in a way, is even more paradoxical than before. It means that if you wait on a floor near the top and fix your attention on any given elevator door, the probability is always high that the next time *that* elevator arrives it will be going up. But the chance that the next elevator to stop on the floor will be going up, regardless of the shaft it is in, is a different matter. *This* probability approaches 1/2 as the number of elevators approaches infinity. The same is true of down elevators stopping on a floor near the bottom.

We assume, of course, that elevators travel independently of one another, with constant speeds, and have the same average waiting time on each floor. If there are just a few elevators, the changes in probability are slight, but if there are 20 or more, the probability gets very close to 1/2 for all floors except the top and bottom ones.

The Bewildered Girlfriends

Have you heard about the boy who never could decide which girlfriend to visit? One girl lived east, the other west. Each day he went to his subway station at random times and took the first train.

Next Train	
Eastbound	Westbound
12:00	12:01
12:10	12:11
12:20	12:21

Both eastbound and westbound trains ran at 10-minute intervals.

One night the east girl said:
Easty: I'm so happy you're able to visit me nine out of ten days on the average.

Another night the west girl got very angry.
Westy: Why is it that I see you only about once every ten days?

Next Train	
Eastbound	Westbound
12:00	12:01
12:10	12:11
12:20	12:21
12:30	12:31
12:40	12:41
12:50	12:51

This curious state of affairs is like that of the elevators. Although all trains run at 10-minute intervals, the schedule is such that a westbound train always arrives and leaves 1 minute after an eastbound train.

To catch a westbound train, the boy must arrive during one of the 1-minute intervals shaded on the clock. To catch an eastbound train he must arrive during one of the 9-minute intervals shown in white. The probability of going west is one-tenth, and the probability of going east is nine-tenths.

In this paradox the waiting times between trains are fixed by the schedule. In a sequence of random events, the "average waiting time" between events is obtained by adding n waiting times and dividing by n. For example, the boy's average waiting time for an eastbound train is 4½ minutes, and his average waiting time for a westbound train is half a minute.

Many other paradoxes involve waiting times. You may enjoy tackling this one. In flipping a coin, the average waiting time for a head (or a tail) is two flips. This means that if you take a long list of coin flips and count the waiting times from each head to the next, the average "run" between heads (not counting the first head, but including the next one) is two flips.

Suppose you have a long vertical list of the outcome of many penny flips. Randomly select a spot between any two adjacent flips (perhaps by closing your eyes and drawing a horizontal line across the list). Find the nearest head above the line and the nearest head below, then count the run from one head to the other. If you do this many times, what will be the average run between heads?

Intuitively, the answer seems to be two flips. Actually it is three. The reason is the same as the reason why the boy usually catches the eastbound train. Some runs between heads will be short, and some long. Your randomly drawn line is like the boy who arrives on the subway platform at random times. It is more likely to strike inside a long run than a short one.

Here is a simple proof that the correct answer is three flips. Pennies have no memories of past behavior, so wherever you draw the line, the average waiting time to the next head must be two. The same applies to the average waiting time if we "time reverse" the process and count backward. Consequently, the average run between heads is twice 2, or 4, if both heads are counted. Since we have defined a run as including one head but not the other, the length of the run is $4 - 1 = 3$.

The comparable problem with a roulette wheel is even more startling. A roulette wheel has 38 numbers including 0 and 00. Thus the average waiting time for a given number, say 7, is 38 spins. But if you take a long list of roulette results, and pick a spot on it at random, the average run that it selects, from one 7 to the next, is not 38 but $(2 \times 38) - 1 = 75$.

Three-Shell Game

Operator: Step right up, folks. See if you can guess which shell the pea is under. Double your money if you win.
After playing the game a while, Mr. Mark decided he couldn't win more than once out of three.

Operator: Don't leave, Mac. I'll give you a break. Pick any shell. I'll turn over an empty one. Then the pea has to be under one of the other two, so your chances of winning go way up.

Poor Mr. Mark went broke fast. He didn't realize that turning an empty shell had no effect on his chances. Do you see why?

After Mr. Mark has selected a shell, at least one of the remaining two shells is certain to be empty. Since the operator knows where he put the pea, he can always turn over an empty shell. Therefore his act of doing so adds no useful information for Mr. Mark to revise his estimate of the probability that he has picked the right shell.

You can demonstrate this with an ace of spades and the two red aces. Mix the cards and deal them face down in a row. Allow someone to put a finger on a card. What is the probability he or she has picked the ace of spades? Clearly it is 1/3.

Suppose, now, you peek at your two cards and turn over a red ace. You can argue, like the shell game operator, as follows. Only two cards are face down. The ace of spades is as likely to be one of them as the other. Therefore the probability that the ace of spades has been picked seems to have gone up to 1/2. Actually, it remains 1/3. Because you can *always* turn over a red ace, turning it adds no new information that affects the probability.

You can puzzle your friends with the following variation. Instead of peeking at the two unselected cards to make sure you turn a red ace, allow the person whose finger is on a card to turn over one of them. If it should be the ace of spades, the deal is declared void and the game is repeated until the reversed card is a red ace. Does this procedure raise the probability of picking the ace of spades?

Oddly enough, it raises it to 1/2. We can see why by taking a sample case. Call the positions of the cards 1,2,3. Let's assume the person puts a finger on card 2, then turns over card 3 and it is a red ace.

There are six equally possible ways the six cards can be dealt:

1. A♠ A♡ A♦
2. A♠ A♦ A♡
3. A♦ A♠ A♡
4. A♦ A♡ A♠
5. A♡ A♠ A♦
6. A♡ A♦ A♠

If the third card (reversed) had been the ace of spades, the deal would have been declared void, therefore cases 4 and 6 never enter into the problem. We rule them out as possible cases. Of the remaining four (1,2,3,5), card 2 (on which the finger rests) is the ace of spades in two cases. Therefore the probability card 2 is the ace of spades is indeed 2/4 = 1/2.

The result is the same regardless of which card the person chooses, and which card is exposed as the red ace. Had Mr. Mark been allowed to pick the shell to be turned over, and had it been empty, his chances of being right *would* have gone from 1/3 to 1/2.

Chuck-A-Luck

The next time you go to a carnival, stay away from chuck-a-luck! Many people are tricked into playing it because they think they can't lose.

With customers thinking this way, it's no wonder casino operators become millionaires! Why does chuck-a-luck give the house a strong percentage?

The chuck-a-luck cage contains three dice, which are shaken by turning the cage. A player bets on any number from 1 to 6 and is paid whatever he bets for each die that shows his number. Players often reason:

Mr. Mark: If the game had one die, my number would show once in six games. If it had two dice, it would show in two out of six games. With three dice, it must show in three out of six games, and that's even odds!

Mr. Mark: But my odds are better than that! If I bet a dollar on, say, 5, and 5 shows on two dice, I win $2. If it shows on all three, I win $3. The game must be in my favor!

Chuck-a-luck is played in many casinos throughout the United States and abroad. In England, back in the early 1800s, it was called Sweat-cloth. In more recent times it has been known as Bird Cage. In British and Australian pubs it is usually played with three dice, each carrying symbols of a spade, a diamond, a heart, a club, a crown, and an anchor, and is called Crown and Anchor.

At carnivals, the operator often shouts as a come-on, "Three winners and three losers every time!" This gives the strong impression of a fair game. The game actually *would* be fair if the dice always showed three different numbers. After each turn of the cage, the operator would collect $3 from the three losers (assuming bets of a dollar each), and pay out $3 to the three winners. Fortunately for the operator, the same number often shows on two or three dice. If it shows on two dice, he takes in $4 and pays out $3, making $1 profit. If it shows on all three dice, he takes in $5 and pays out $3, making a profit of $2. These doublets and triplets provide the house its percentage.

Calculating the house percentage by formulas is a tricky business. The safest way is to make a full listing of the 216 ways three dice can fall. You will find that only 120 of them show all three dice different, 90 show doublets, and 6 show all three dice alike. Assume that the game is played 216 times, with the 216 possible outcomes. For each game six people each bet a dollar on each of the six numbers. The operator will collect a total of 216 × $6 = $1296 in bets.

When all three dice are different, he pays back a total of 120 × $6 = $720. When doublets show he pays out a total of 90 × $2 = $180 for the singlet, and 90 × $3 = $270 for the doublets. When triplets show he pays out 6 × $4 = $24. This is a total payoff of $1194, giving him a profit of $102. Dividing $102 by $1296 gives the house percentage of 7.8+ percent. This means that for every dollar a player wagers, he can expect in the long run to lose about 7.8 cents.

What are the chances of winning on a single throw? If the three dice are colored red, green, and blue, there are 36 ways the red die can show a 1 while the other dice come up as they please. Continuing our count there are 30 ways the red die can show something other than 1, while the green die shows a 1 and the blue die comes up as it pleases. Finally, if the blue die shows a 1 and the red and green dice can show anything except a 1, there are 25 cases. Thus, in 91 cases out of 216 at least one die shows a 1. So the probability of winning on 1 is 91/216, or considerably less than 1/2, and the same is true for any other number.

Puzzling Parrots

A lady owned two parrots. One day a visitor asked:
Visitor: Is one bird a male?
Owner: Yes.
What is the probability both birds are males? It's one-third.

Suppose the visitor asks:
Visitor: Is the dark bird a male?
Owner: Yes.
Now the probability both birds are males goes up to one-half. This doesn't make sense. Why does asking about the dark bird change the probability?

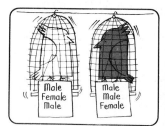

This paradox can easily be explained by listing all the possible cases. When the visitor knows that one bird is a male, there are just three cases to consider. Only one is male–male, so the probability that both are males is one-third. (We assume equal likelihood that each parrot is male or female.)

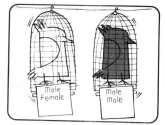

But when the visitor knows that the dark bird is male, there are now just two cases to consider. Only one is male–male, so the probability both are male is one-half.

You can model the parrot problem by having someone toss two coins, one a penny, the other a nickel, then make certain statements about the result. The tosser can adopt one of several procedures:

1. If both coins are heads he says: "At least one coin is a head." If both are tails, he says: "At least one coin is a tail." If the coins are different, he says: "At least one coin is a ---," picking heads or tails at random. What is the probability that both coins show whatever side was called? Answer: 1/2.

2. The tosser has agreed in advance to call out "At least one coin is heads" only when this is the case. If no coin is heads, he says nothing and tosses again. What is the probability both coins are heads? Answer: 1/3 (because now the possibility of tails–tails is eliminated.)

3. The tosser agrees in advance to call out how the penny fell, regardless of whether it is heads or tails. What is the probability that the coins match? Answer: 1/2.

4. The tosser agrees in advance to call out "At least one coin is heads" only when the penny is heads. What is the probability both are heads? Answer: 1/2.

The parrot paradox is sometimes given in such an ambiguous way that it is not possible to answer it. For example, suppose you meet a stranger who says: "I have two children. At least one is a boy." What is the probability both children are boys?

This is not a precisely defined problem because you know nothing about the circumstances that prompted the man to make his statement. He might just as likely have said "At least one is a girl," picking the sex at random if his children are different sexes, and naming the sex if both are the same. If *that* was his procedure, the probability of both being boys is 1/2. The situation corresponds to No. 1 above.

In our parrot problem, ambiguity is eliminated by having the customer ask the question. The first question, "Is at least one bird a male?" corresponds to No. 2 above. The second question, "Is the dark bird a male?" corresponds to No. 4.

An even more astounding paradox, closely related to the two parrots, is known as the paradox of the second ace. Assume you are playing bridge. After the cards are dealt you look over your hand and announce: "I have an ace." What is the probability you have a second ace? It is exactly 5359/14498, which is less than 1/2.

Suppose now, that all of you agree on a particular ace, say the ace of spades. The play continues until you are dealt a hand that enables you to declare: "I have the ace of spades." What is the probability you have a second ace? It is now 11686/20825, or slightly *better* than 1/2! Why should naming the ace alter the odds?

The computation of chances for the entire deck is long and tedious, but the structure of the paradox can be understood easily by reducing the deck to four cards, say the ace of spades, ace of hearts, two of clubs, and jack of diamonds.

(Simplifying a problem by reducing its number of elements is often an excellent way to understand its structure.) The four-card deck is shuffled and dealt to two players. There are six equally probable two-card hands:

A♠ A♡
A♠ J♢
A♠ 2♣
A♡ J♢
A♡ 2♣
J♢ 2♣

Five of the six hands permit a player to say "I have an ace," but in only one of the five hands is there a second ace. Consequently the probability of the second ace is 1/5.

There are just three hands that permit the player to declare "I have the ace of spades." Only one of the three has a second ace. Therefore the probability of the second ace is 1/3.

Note that the ace to be named must be agreed upon in advance, as well as the person who, in each case, makes the announcement that he or she holds an ace. If these assumptions are not explicitly made, the problem is not precisely defined.

The Wallet Game

Professor Smith is having lunch with two math students.

Professor Smith: Let me show you a new game. Put your wallets on the table. We'll count the money in each. Whoever has the smallest amount wins all the money in the other wallet.

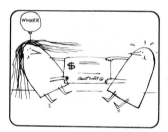

Joe: Hmm. If I have more than Jill, she'll win just what I have. But if she has more, I'll win more than I have. So I'll win more than I can lose. The game must be in *my* favor.

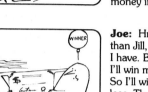

Jill: If I have more than Joe, he'll win just what I have. But if he has more, I'll win, and I'll win more than I have. The game's in *my* favor.

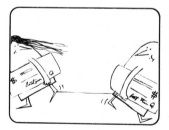

How can a game be favorable to both players? It can't. Does this paradox arise because each player wrongly assumes his chances of winning or losing are equal?

This charming paradox comes from the French mathematician Maurice Kraitchik. In his book *Mathematical Recreations* he gives it with neckties instead of wallets:

> Each of two persons claims to have the finer necktie. They call in a third person who must make a decision. The winner must give his necktie to the loser as consolation. Each of the contestants reasons as follows: "I know what my tie is worth. I may lose it, but I may also win a better one, so the game is to my advantage." How can the game be to the advantage of both?

If we define the situation precisely by making certain assumptions, it is a fair game. Of course, if we have information that one player habitually carries less money (or wears a cheaper necktie) than the other, then we know it is not a fair game. If no such information is available, we may assume that each player has a random amount of money from zero to any specified amount, say $100. If we construct a payoff matrix on this assumption, as Kraitchik does in his book, we see that the game is "symmetrical" and does not favor either player.

Unfortunately, this does not tell us what is wrong with the reasoning of the two players. We have been unable to find a way to make this clear in any simple manner. Kraitchik is no help, and so far as we know, there is no other reference on the game.

The Principle of Indifference

Is there life on Titan, the largest moon of Saturn?

Will there be an atomic war?

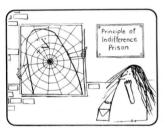

If you answer questions like these by saying that yes and no are equally probable, you are foolishly applying what is called the "principle of indifference." Careless use of this principle has caught many mathematicians, scientists, and even great philosophers in webs of absurdity.

The "principle of insufficient reason," which the economist John Maynard Keynes renamed the "principle of indifference" in his famous *Treatise on Probability,* can be stated as follows: If we have no good reasons for supposing something to be true or false, we assign even odds to the probability of each truth value.

The principle has had a long and notorious history, with applications in such diverse fields as science, ethics, statistics, economics, philosophy, and psychic research. If not properly used, it leads to absurd paradoxes and outright logical contradictions. The French astronomer and mathematician Laplace once used the principle as a basis for calculating that the probability of the sun rising tomorrow is 1,826,214 to 1!

Let's see how contradictions arise if the principle is carelessly applied to our questions about Titan and atomic war. What is the probability there is some form of life on Titan? We apply the principle of indifference and answer 1/2. What is the probability of *no* simple plant life on Titan? Again, we answer 1/2. Of no one-celled animal life? Again, 1/2. What is the probability there is neither simple plant life nor simple animal life on Titan? By the laws of probability we must multiply 1/2 by 1/2 and answer 1/4. This means that the probability of *some* form of life on Titan has now risen to $1 - 1/4 = 3/4$, contradicting our former estimate of 1/2.

What is the probability of an atomic war before the year 2000? By the principle of indifference we reply 1/2. What is the probability of no atom bomb dropped on the United States? Answer: 1/2. Of no atom bomb on Russia? Answer: 1/2. Of no atom bomb on France? Answer: 1/2. If we apply this reasoning to ten different countries, the probability of no atom bomb falling on any of them is the tenth power of 1/2, or 1/1024. Subtracting this from 1 gives us the probability that an atom bomb will fall on one of the ten countries—a probability of 1023/1024.

In both of the above examples the principle of indifference is aided by an additional assumption in yielding such absurd results. We have tacitly assumed the independence of events that clearly are not independent. In light of the theory of evolution, the probability of intelligent life on Titan is dependent on the existence there of lower forms of life. Given the world situation as it is, the probability of an atom bomb falling on, say, the United States is not independent of the probability of such a bomb falling on Russia.

Another good example of careless use of the principle of indifference is the paradox of the unknown cube. Suppose you are told that a cube, hidden in a closet, has a side that is between 2 and 4 feet. You have no reason to assume the side is less than 3 or more than 3, so your best guess of the cube's side is 3. Now consider the cube's volume. It must be between $2^3 = 8$ and $4^3 = 64$ cubic feet. You have no reason to think the volume is less than 36 or more than 36, so you guess the volume to be 36 cubic feet. In other words, your best estimate is a cube with a side of 3 and a volume of 36, which would be a queer sort of cube! Put another way, if you apply the principle of indifference to the cube's side, you get a cube of side 3 and volume 27. Apply it to the volume, and you get a cube of volume 36 and a side equal to the cube root of 36, or about 3.30 feet.

The cube paradox is a good model for showing the kind of trouble a scientist or statistician can get into when he or she obtains minimum and maximum values for a quantity, then assumes that the actual value is most likely to be halfway between. Many other examples of such paradoxes are given in Keynes' book.

The principle has legitimate applications in probability, but only when the symmetries of a situation provide objective grounds for assuming probabilities to be equal. For example, a penny is geometrically symmetrical in the sense that you can pass a plane of symmetry edgewise through the coin. It is physically symmetrical in having a uniform density; that is, it is not weighted on one side. The forces that act upon it in the air—gravity, friction, air pressure, and so on—are symmetrical, not favoring one side over the other. We are justified, therefore, in assuming that heads and tails have equal probability. Similar symmetries apply to the six sides of a cubical die, or the 38 slots of a roulette wheel. In each of these cases extensive experiments in gambling casinos around the world have shown the correctness and the limitations of these symmetry assumptions. In cases where such symmetries are not known, or may not even exist, an application of the principle of indifference often leads to absurd results.

Pascal's Wager

Blaise Pascal, a famous seventeenth century French mathematician, applied the principle of indifference to Christian faith.

Pascal: A person cannot decide whether to accept or reject the doctrines of the church. They may be true. They may be false. It is like the flip of a coin. The odds are even. But what are the payoffs?

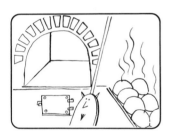

Pascal: Suppose you reject the church. If the church is false, you lose nothing. But if the church is true, you face infinite suffering in hell.

Pascal: Suppose you accept the church. If the church is false, you gain nothing. But if it is true, you win eternal bliss in heaven.

Pascal was sure that the payoffs of this decision game are infinitely in favor of a bet that the church is true. Philosophers have been debating Pascal's wager ever since. What's your opinion?

Blaise Pascal was one of the founders of probability theory. In the first picture he is pointing to a famous pattern of numbers called "Pascal's triangle." Pascal did not invent the triangle (it goes back to the early Middle Ages), but he was the first to make a thorough investigation of it. The pattern has elegant combinatorial properties that make it a useful tool in answering elementary problems in probability. (See the chapter on Pascal's triangle in Harold Jacobs, *Mathematics: A Human Endeavor.*)

Pascal's argument for becoming a Christian, or "Pascal's wager" as it is usually called, is given in Thought 233 of his *Pensees*. The wager suggests many provocative questions. For instance:

1. Is the principle of indifference legitimately applied in Pascal's argument?

2. How would you answer this objection by the French philosopher Denis Diderot? There are many other great religions, such as Islam, that also make salvation conditional on acceptance of the religion. Does Pascal's wager apply to all of them? If so, should one become a member of each?

3. What do you think of William James' watered-down version of the wager? In his essay "The Will to Believe," James argued that a decision to believe in God (James was not concerned with an afterlife or a particular church) is a good bet because there is no evidence one way or another concerning God's existence, therefore one should make whatever decision makes him the happiest throughout his life.

4. What do you think of this argument by H. G. Wells? We do not know whether the world will or will not survive an atomic holocaust. But you should live and behave as if you are sure it will survive because, as Wells put it, "if at the end your cheerfulness is not justified, at any rate you will have been cheerful."

Statistics

Paradoxes about gismos, clumps, ravens, and grue

Statistics—which concerns the obtaining, organizing, and analyzing of numerical information—is increasingly important in today's highly complex world. Average citizens are so bombarded by data, from the state of the economy to the effectiveness of brands of toothpaste, that unless they have some knowledge of elementary statistics they are incapable of making intelligent decisions. It would be hard to find a science in which statistical studies do not play vital roles, not to mention the indispensability of statistics in dozens of other fields—insurance, public health, advertising—and almost every type of business.

In no sense is this chapter an introduction to statistics. Taken alone, it will not teach you the basics of the field. What it does try to do is offer a sampling of colorful paradoxes that will stimulate your interest in learning more about the underlying mathematics.

The chapter opens with a story that introduces those three famous fundamental measures: the mean, the median, and the mode. This is followed by some outlandish examples of the misuse of data—the great art of "lying" with statistics—that will alert you to some common pitfalls.

Faced with today's explosion of interest in astrology and all things paranormal, few people are aware of how a lack of statistical sophistication makes it easy for them to be impressed by surprising coincidences, which in the light of probability theory and statistics are not at all surprising.

Consider, for example, the notorious birthday paradox. Among any randomly selected group of 23 people, the chances are slightly better than 1/2 that at least two will have birthdates of the same month and day! If there are 40 people the chances of such a coincidence rise to about 9/10.

One's first reaction is total disbelief. Next, one makes an empirical test at a party of some 40 guests, or by checking 40 names at random in a *Who's Who*. The third step, if you have any curiosity about the mathematics behind this paradox, is to learn enough about it to understand why things turn out this way. It is in just this way that these paradoxes provide marvelous stepping stones to significant mathematics.

Instructions are given for some card tricks in which seemingly miraculous coincidences are the natural outcome of simple mathematical laws. The voting paradox is one of the most famous of many strongly counterintuitive theorems studied in decision theory, a new branch of mathematics concerned with making rational decisions on the basis of statistical information. A story about Miss Lonelyhearts dramatizes another astonishing, little-known paradox.

The chapter concludes with two of the most widely debated paradoxes in the philosophy of science: the notorious paradox of the raven and a paradox about a strange property called "grue." Both point up the importance of statistics in evaluating the degree of credibility of scientific hypotheses.

The Deceptive "Average"

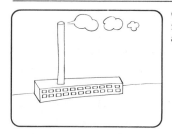

Gismo Products has a small factory where supergismos are manufactured.

The management consists of Mr. Gismo, his brother, and six relatives. The work force consists of five foremen and ten workers. Business is good, and the factory needs a new worker.

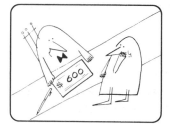

Mr. Gismo is interviewing Sam for the job.
Mr. Gismo: We pay very well here. The average salary is $600 a week. During your training period you'll get $150 a week, but that will soon increase.

After working a few days, Sam asked to see the boss.
Sam: You misled me! I've checked with the other workers and not one is getting more than $200 a week. How can the average salary be $600 a week?

Mr. Gismo: Now, Sam, don't get excited. The average salary is $600. I'll prove it to you.

Mr. Gismo	$4800
Mr. Gismo's Brother	$2000
6 Relatives ($500/wk)	$3000
5 Foremen ($400/wk)	$2000
10 Workers ($200/wk)	$2000
	$13,800

Mr. Gismo: Here's what we pay out each week. I get $4800, my brother gets $2000, my six relatives each make $500, the five foremen each make $400, and the ten workers each get $200. That makes a weekly total of $13,800 for 23 people, right?

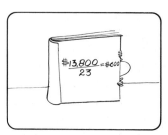

$$\frac{\$13,800}{23} = \$600$$

Sam: Okay, okay. You're right. The average is $600 a week. But you *still* misled me.

Median $400

Mr. Gismo: I disagree. You just didn't understand. I could have listed the salaries in order and told you that the middle salary is $400, but that isn't the average. It's the *median*.

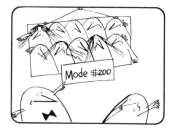

Mode $200

Sam: Where does the $200 a week come in?
Mr. Gismo: That's called the *mode*. It's the salary that *most* people are making.

Average
Mode
Median

Mr. Gismo: The trouble with you, my boy, is that you don't know the difference between average, median, and mode.
Sam: Well, I know now. And . . . I quit!

Statistical statements can be extremely paradoxical and at times downright deceptive. The story about Mr. Gismo's factory brings out a common source of misunderstanding about the differences between the mean, the median, and the mode.

The word *average* is usually an abbreviation for "arithmetic mean." It is a valuable statistical measure. If, however, there are a few extreme values, such as the high salaries of the top two men in Mr. Gismo's factory, the "average" salary can convey a false impression.

It is easy to find similar situations in which a statement about the "average" is equally misleading. For example, a newspaper reports that a man has drowned in a river that has an average depth of only 2 feet. Is this surprising? Not when you learn that he drowned in one of the few spots where the depth is more than 10 feet.

A corporation may report that its policies are democratically controlled by stockholders because its fifty stockholders have altogether 600 votes, or an average of 12 votes per person. If, however, forty-five stockholders have only 4 votes each, and five persons have 84 votes each, the average is still 12 votes per person, but five people have complete control of the corporation.

One more example: To attract retail businesses, the Chamber of Commerce boasts that the town's average per capita income is very high. Most people would take this to imply that a majority of the town's citizens are in high-income brackets. But if one billionaire happens to live in the town, the other residents could all have low incomes and the "average" per capita income would still be high.

The reporting of statistics is made even more confusing by the fact that the word *average* is sometimes used, not for an arithmetic mean, but for the median or the mode. The median is the value in the middle of a list of values arranged in order of magnitude. If there is an odd number of items on the list, the median is simply the middle value. If there is an even number, the median is customarily taken as the arithmetic mean of the two values in the middle.

The median is a more useful measure to Sam than the arithmetic mean, but even the median gives a distorted picture of the firm's salaries. What Sam really needed to know was the mode—the value that appears most often on a list. In this case, the mode is the salary paid to more people in the firm than any other salary. It is sometimes called the "typical case" because it occurs more often than any other. In our last example, a "typical" family in a town—one that represents the income mode—may be very poor, even though the town's average income, due to a small number of very wealthy people, is very high.

Mother of the Year

Later in the year Sam's wife received an award from the town's mayor. She had been named the mother of the year.

The local paper ran a picture of Sam, his wife, and their 13 children.

The editor was pleased with the photograph.
Editor: Good work, Bascom. I have a new assignment. Get me a picture of the average-sized family in this town.

Bascom was unable to do this. Why? Because not a single family in town was average! The computed average number of children was 2½.

Another common misconception about an "average" is that actual instances of the average must exist. After seeing this episode, in which we learn that there *is* no average family of 2½ children, you should have no difficulty thinking of other instances in which the average value is not represented by any individual case. Can you, for example, toss a die and obtain the average number of spots a tossed die shows in the long run?

Here are some other questions to ask yourself to sharpen your understanding of the arithmetic mean, median, and mode.

1. If the editor wants a photograph of a "typical" family, in the sense of the mode, can the photographer always find such a family? (Yes, the typical case obviously exists).

2. Is it possible that there is more than one mode? For example, could a family of two children and a family of three children each be examples of a mode? (Yes, if the town had 1,476 families with two children, and 1,476 families with three children, and all other families had more or less children, then the town would have two types of families, each a legitimate mode.)

3. If the editor wants a photograph of a median family, can he always find one? (Usually, but not always. As we saw above, if there is an even number of families in a town, and the two middle families are not alike in the number of children, then the median need not be an integer.)

Jumping to Conclusions

 Statistics show that most car accidents occur when cars travel at moderate speeds, and that very few accidents occur at speeds of more than 150 kilometers per hour. Does this mean it is safer to drive at higher speeds?

 A research study showed that children with big feet could spell better than those with small feet. Does this mean that the size of one's foot is a measure of one's ability to spell?

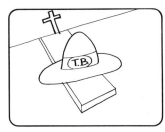 Not at all. Statistical relationships often have nothing to do with cause and effect. Most people drive at moderate speeds, so naturally most accidents occur at moderate speeds.

 It does not. The study included *growing* children. All it showed was that older children, who, of course, have bigger feet, spell better than younger ones.

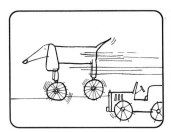 If statistics showed that more people died of tuberculosis in Arizona than in any other state, would that mean that Arizona's climate favored getting TB?

 Quite the contrary. Arizona's climate is so *helpful* to TB victims that thousands of them go there. Naturally this would raise the average number of TB deaths.

These three episodes underscore the importance of not jumping to conclusions about cause and effect when you hear of a statistical correlation. Here are a few more instances:

1. It is often said that most car accidents occur near the home. Does this mean that travel on highways, many miles from home, is safer than driving around town? No. The statistics simply reflect the fact that a car is more often driven near one's home than on distant highways.

2. A study showed that a certain state ranked high both in the proportion of people who drink milk and in the proportion of people who die of cancer. Does this indicate that drinking milk causes cancer? No. The state also ranked high in the proportion of people who are elderly. Since cancer is commonly an affliction of the aged, this is what raised the proportion of cancer deaths.

3. A study showed that in a certain city there was a sharp rise both in deaths from heart failure and in the consumption of beer. Could it be that beer drinking increases the probability of a heart attack? No. The increase in both cases was the result of a rapid rise in population. By the same reasoning, heart attacks could be attributed to hundreds of other things: increased consumption of coffee, increased chewing of gum, increased playing of bridge, increased television watching.

4. A study showed that at the same time a certain European city had a large increase in population, there was a large increase in the number of stork nests in the city. Does this support the belief that storks bring babies? No, it reflected the fact that as the number of buildings increased, there were more places in the city where storks could nest.

5. A recent study showed that most great mathematicians were eldest sons. Does this mean that first-born sons are more likely to have mathematical ability than sons born later? No, it simply reflects the surprising fact that most *sons* are eldest sons.

The last example suggests some interesting experiments. Survey your male friends to see if more than half are eldest sons. Do the same with your female friends to see what proportion are eldest daughters.

Or a thought experiment. Consider a population in which 100 families each have two children. What fraction of the boys (or girls) will be eldest sons (or daughters)? (Answer: 3/4.) Compute the fraction in a population of 100 families, each with three children. (Answer: 7/12.) It goes without saying that in families of only one child, that child will be the eldest.

The exact percentage of eldest children of one sex obviously will vary with the sizes of families in the population under consideration, but for all populations it is over 1/2 and for most populations the figure is well over 1/2.

These examples may stimulate you to look for other instances of statistical statements that are easily misinterpreted with respect to cause and effect. Modern advertising, especially television commercials, is a rich source of such misleading assertions.

The Small-World Paradox

Many people these days believe that coincidences are caused by stars or other occult forces.

For example, suppose two strangers meet on an airplane.

Jim: So you're from Boston! My old friend Lucy Jones is an attorney there.

Tom: What a small world! She's my wife's best friend!

Is this sort of thing an unlikely coincidence? Statisticians have proved it is not.

Most people are very surprised when they meet a stranger, especially if far from home, and discover that they have a friend in common. A group of social scientists at MIT, led by Ithiel de Sola Pool, made a study of this "small-world paradox." They found that if two people in the United States are selected at random, on the average each person will know about 1000 people. This gives a probability of about 1 in 100,000 that they will know each other. The probability rises sharply to about 1 in 100 that they have one friend in common. The probability that they are connected by a chain of *two* intermediates (as in the dialog at left) is actually better than 99 in 100! In other words, if Brown and Smith are two persons in the United States picked at random, the chances are almost certain that Brown will know someone who knows someone who knows Smith.

Psychologist Stanley Milgram approached the small-world problem by selecting a random group of "starting persons." Each was given a document to transmit to a "target person" (unknown to the starting person) who lived in a distant state. This was to be done by mailing the document to a friend (someone known on a first-name basis) who seemed most likely to know the target person, and the friend in turn would mail it to another friend, until finally it reached someone who knew the target person. Milgram found that the number of intermediate links, before the document reached the target person, varied from 2 to 10, with the median at 5. When people were asked how many intermediate links would probably be necessary, most of them guessed about 100.

Milgram's study shows how tightly people are joined by a network of mutual friends. Thus, it is not surprising that two strangers, meeting far from home, will have a mutual acquaintance. The network also explains other unusual statistical phenomena, such as the speed with which gossip, sensational news, confidential information, and jokes are transmitted.

What's Your Sign?

These four people are meeting for the first time. Would it be a remarkable coincidence if at least two have the same astrological sign?

You might think so, but actually it will happen about four times out of ten. Assume that each of the four people could have been born, with equal probability, under any 1 of the 12 signs of the zodiac. What is the probability that at least two of the people will have the same sign?

Let's model the problem with a deck of cards. Remove all the kings. The deck will then consist of 12 values for each of the four suits. Let each suit represent a person, and each value represent a zodiac sign. If we select at random a card from each suit, what is the probability that at least two cards will match in value? This will be the same as the probability that at least two of the four strangers will have identical astrological signs.

The simplest way to solve this problem is to calculate the probability that *no* two card values are alike. Subtracting this from 1 will give the probability we seek.

If we consider two suits, say hearts and spades, the probability of no match is 11/12 because there is only one chance in 12 that a heart and a spade will have the same value. The probability that a club will differ in value from the other two cards is 10/12, and the probability that a diamond will differ from the other three cards is 9/12. The product of these three fractions gives us the probability that *none* of the four cards match. It is 55/96. Subtracting this from 1 gives 41/96, or a probability of about 4/10 that at least two of the four persons will have the same sign. This is almost 1/2, so such a coincidence is not surprising.

This is a variation of the well-known birthdate paradox. If 23 people meet at random, the probability is a trifle better than 1/2 that at least two will be born on the same day of the same month. The calculation proceeds as before, except that now we have 22 fractions to be multiplied:

$$\frac{364}{365} \times \frac{363}{365} \times \frac{362}{365} \times \cdots \times \frac{343}{365}$$

The probability is 1 minus the product, or $0.5073+$, a trifle better than 1/2. Confirming this figure is easy with a pocket calculator. The probability of a matching birthdate rises rapidly if there are more than 23 persons. Among 30 people the probability is about 7/10 that at least two will have identical birthdates. Among 100 persons the odds are higher than 3 million to 1.

Some questions you might think about:

1. How many presidents of the United States have the same birthdate? How many have the same date of death? How do these results compare with theoretical expectation?

2. What is the smallest number of people for whom the probability is better than 1/2 that at least two are born in the same month? (Answer: 5. The probability of a match is 89/144 or about 0.62.)

3. What is the smallest number of people for whom the probability is better than 1/2 that at least two are born on the same day of the week? (Answer: 4. The probability of a match is 223/343 or about 0.65.)

4. What is the smallest number of people for whom the probability is better than 1/2 that at least one of them has the same birthdate as *yours*? (Answer: 253. It is *not* 183, as it would be if everyone had a birthdate unlike anyone else.)

Patterns in Pi

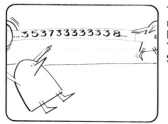

The digits of π seem random, but look what happens starting at the 710,100th decimal digit. Seven 3's in a row!

The digits of π are not random in the sense of being randomly generated, but they are "random" in the sense that they are patternless. Mathematicians have subjected the decimal expansion of π to all sorts of tests to discover any "order" in the digits, but without success. In this sense π is as disordered as digits obtained by a spinner that can stop at any digit from 0 to 9.

The actual odds against a run of seven 3's in π, starting at any given random spot in its decimal expansion, are high. The odds are 9,999,995 to 1 against such a run. Therefore, that this run occurs in the first 710,106 decimal digits of π may at first seem surprising. But if we are searching π for *any* sort of unusual pattern of 7 digits, the probability of finding such a pattern goes up. Scores of other patterns would have been equally surprising: 4444444, or 8888888, or 1212121, or 1234567, or 7654321. Since we do not know *in advance* what sort of pattern we are looking for, it is a good bet that we can discover *some* kind of unusual pattern. The only limit is our ingenuity in searching for such patterns. As Aristotle once put it, the improbable is *extremely probable*.

Jason and the Sun

This man has written the initial letters of the months: J for January, F for February, and so on. Is the word JASON a coincidence?

Here are the first letters of the nine planets according to their distances from the sun. M for Mercury, V for Venus, and so on. Is the word SUN another coincidence?

These two amusing coincidences underscore the truth of Aristotle's dictum. Another way to demonstrate the probability of the improbable is with a spinner that randomly selects a letter of the alphabet. If you pick a word of, say, three letters and bet that it will appear as a consecutive sequence of letters obtained by 100 spins, it would be an unfavorable bet. But if you bet on the appearance of just *any* 3-letter dictionary word, it would be a favorable bet.

You can use such a spinner to select letters, writing them down one at a time and seeing how long it takes until three consecutive letters form a recognizable 3-letter word. Try for 4- and 5-letter words. It is surprising how often such words occur.

A dramatic and weird touch can be added by considering ways in which each word that you get can be related to current events. "Eva," for instance, may be the name of someone you know, or the word "hat" may remind you of someone who lost a hat. Watch for combinations (FBI, IBM, USA), abbreviations (Fla, Dec, Fri), and initials (FDR, JFK). Connecting such "words" with events is so easy that you can easily see how someone might believe occult influences are at work in the formation of these words!

The experiment explains why so many remarkable coincidences occur in one's lifetime. When they happen, there is a strong tendency to believe that mysterious forces are at work. To a statistician, such coincidences are extremely probable. There are millions and millions of ways that a coincidence of *some* sort can arise in the multitude of events that occur every day. Since the nature of the coincidence is not specified in advance, it is like the unspecified pattern of digits in π or the unspecified word that turns up when letters are picked at random. When the coincidence takes place, it always seems too improbable to have occurred by chance. What we forget is that for every such coincidence, billions of other possible coincidences that *might* have occurred didn't.

Crazy Clumps

Even a shuffled deck of cards will contain coincidences. For instance, almost always there will be a clump of six or seven cards of the same color.

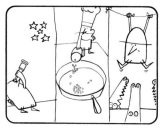

Stars clump in groups called constellations. Beans tossed on a surface tend to form little clumps. There is an old saying: "Bad luck comes in threes."

The tendency of random events to "clump" in various ways is a well-recognized phenomenon, and entire books have been written about what statisticians call *clumping theory*. The run of seven 3's in π is an example of random clumping. If you keep flipping a penny, or spinning a roulette wheel and recording the colors and numbers, you will find similar examples of long runs turning up with surprising frequency.

A striking experiment in clumping was discovered by A. D. Moore, an engineer at the University of Michigan. Moore calls it the "nonpareil mosaic" because it uses large quantities of nonpareils, a sugar candy manufactured in the shape of tiny colored spheres. Obtain enough red and enough green nonpareils so that you can fill a glass bottle with equal amounts of each. Shake the bottle until the two colors are thoroughly mixed.

Inspect the sides of the bottle. You would expect to see a homogeneous mix of colors, but instead you see a beautiful mosaic made up of irregular large red clumps interspersed with equally large green clumps. The pattern is so unexpected that even mathematicians, when they first see it, believe that some sort of electrostatic effect is causing spheres of like color to stick to one another. Actually, nothing but chance is operating. The mosaic is the normal result of random clumping.

If this seems hard to believe, try this simple experiment. On a sheet of graph paper, outline a 20-by-20 square. Take each cell in turn and color it red or green, choosing the color by flipping a coin. When the 400-cell square is fully colored, you will see the same kind of mosaic that appeared on the sides of the bottle.

Nonmathematical factors often do enter into clumping. If cars were spaced at random along a freeway and observed from a helicopter, they would appear in clumps, but their actual clumping is much greater than can be explained by chance because drivers tend not to pass cars moving at about the same speed as they are moving and to speed up when there are long open spaces ahead. The positions of towns on a map, the sequences of rainy days, patches of clover and crabgrass on a lawn, and endless other things provide instances of clumping that *exceed* what is caused by chance.

An Amazing Card Trick

Here's an astonishing card paradox related to clumping theory. First arrange a deck so that the colors alternate.

Then cut the deck into two parts, making sure the bottom cards are opposite colors.

Now shuffle one half into the other with one thorough riffle shuffle.

Take the cards from the top in pairs. In spite of the shuffle, *every* pair will be red and black!

This remarkable card trick is an example of how a concealed mathematical structure will enter into clumping and produce a result that seems miraculous. Magicians know it as the Gilbreath principle after Norman Gilbreath, a mathematician and amateur magician, who discovered it in 1958. Since then, hundreds of clever card tricks have been based on it.

Here is an informal proof by mathematical induction of why it works. The deck is cut so that the bottom cards of each half are different. After the first card falls from a thumb to the table, the bottom cards of the two halves will be the *same* color, and opposite the color of the card that fell. It makes no difference, therefore, which of these two cards falls next. In either case, a card of opposite color will fall on top of the first card, producing a pair on the table that do not match. The situation is now exactly the same as before. The bottom cards of the two halves do not match. Whichever card falls, both the remaining bottom cards will be the same color, so irrespective of which card falls next, a second pair of nonmatching cards goes to the table. And so on for all the rest of the cards in the deck.

A good way to present the trick to friends is secretly to prearrange the deck so that the colors alternate. Ask someone to deal cards from the top of the deck onto a pile until the pile contains about 26 cards. (This is a way of making sure that the bottom cards of the two portions do not match.) Let him riffle shuffle the two halves together. Hold this "shuffled" deck under a table so that no one, including yourself, can see the cards. Tell your audience you can "feel" the colors with your fingers, and that you will take the cards from the deck in pairs so that each pair is a red and black card. All you need do, of course, is simply take the pairs from the top of the deck.

Can this remarkable principle be generalized to produce other magic tricks? Try the following procedure. Arrange the deck in a sequence of suits, such as SHCD, SHCD, SHCD, and so on. Deal from the top to form a pile of about 26 cards (the exact number does not matter!). This dealing automatically reverses the order of the cards. Now riffle shuffle the two halves together. Take the cards from the top in quadruplets. Each set of four will contain one card of each suit!

For another surprise, arrange the deck in four sets of 13 cards each, with the cards of each set in the order of Ace, 2, 3, 4, 5, 6, 7, 8, 9, 10, jack, queen, king, without regard to suits. Follow the same procedure of dealing and shuffling as before. Take the cards from the top in sets of 13. Each will contain one card of each value!

For the ultimate generalization, arrange two decks so the order of cards in one deck exactly matches the order of cards in the other. Put one deck on top of the other. Deal off the top to form a pile of about 52 cards. Shuffle the two "decks," then divide the 104 cards exactly in half. Each half will be a complete deck!

The Voting Paradox

Suppose three persons—Abel, Burns, and Clark—are running for president.

Mr. Clark: Two-thirds of the voters like me better than Abel!

A poll shows that 2/3 of the voters prefer A to B, and 2/3 prefer B to C. Will most voters prefer A to C?

Not necessarily! If voters rank the candidates as shown, a startling paradox arises. Let's let the candidates explain it.

Mr. Abel: Two-thirds of the voters like me better than Burns.

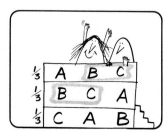

Miss Burns: Two-thirds of the voters like me better than Clark.

This paradox, which goes back to the eighteenth century, is a famous example of a nontransitive relation that can arise when people make pairwise choices. The concept of transitivity applies to such relations as "taller than," "greater than," "less than," "equals," "earlier than," and "heavier than." In general, when a relation R that holds for xRy and yRz also holds for xRz, the relation is said to be transitive.

The voting paradox boggles the mind because we expect the relation *prefers* always to be transitive. If someone prefers A to B, and B to C, we naturally expect him or her to prefer A to C. The paradox shows that this is not always the case. A majority of voters prefer candidate A to B, a majority prefer B to C, and a majority prefer C to A. The situation is nontransitive! The paradox is sometimes called the Arrow paradox after the Nobel-Prize-winning economist Kenneth J. Arrow, who showed from this and other logical considerations that a perfect democratic voting system is in principle impossible.

The paradox can arise in any situation in which a decision must be made between three alternatives that are ranked pairwise with respect to three properties. Suppose A, B, and C are three men who have proposed marriage to the same woman. The rows of a matrix can be interpreted to show how she ranks the three men with three traits, such as intelligence, good looks, and income. Taken by pairs, the woman may find that she prefers A to B, B to C, and C to A!

Mathematician Paul Halmos proposed letting A, B, and C stand for apple, blueberry, and cherry pie. A restaurant offers only two of them at one time. Matrix rows show how a customer ranks the pies with respect to taste, freshness, and size of slice. It is perfectly rational for such a customer to prefer apple to blueberry, blueberry to cherry, and cherry to apple!

For more on nontransitive paradoxes see the following articles in *Scientific American:* my Mathematical Games column (October 1974), "The Choice of Voting Systems" by Richard G. Niemi and William H. Riker (June 1976), and Lynn Steen's Mathematical Games column on voting systems (October 1980).

Miss Lonelyhearts

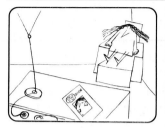

Miss Lonelyhearts, a statistician, is tired of sitting home alone.
Miss Lonelyhearts: I wish I knew some men who weren't married. I think I'll join a group for single people.

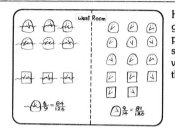

Her statistics for the West group were similar. The proportion of mustached swingers was 84/126. This was greater than 81/126 for the cleanshaven swingers.

Miss Lonelyhearts joined *two* such groups. One evening both groups had parties at Club Paradox. One group met in the East Room, the other in the West Room.

Miss Lonelyhearts: How simple! At *both* parties, I'll have a better chance to meet a swinger if I look for men with mustaches.

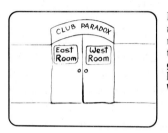

Miss Lonelyhearts: Some men have mustaches and some don't. Some men are swingers and some are squares. I'd like to meet a swinger tonight. Should I look for a man with a mustache?

By the time Miss Lonelyhearts got to Club Paradox, the two groups had decided to combine. Everybody had moved to the North Room.

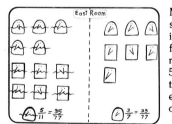

Miss Lonelyhearts made a statistical study of the men in the East group. She found that the proportion of mustached swingers was 5/11 or 35/77. The proportion of cleanshaven swingers was smaller. It was 3/7 or 33/77.

Miss Lonelyhearts: What shall I do now? If a mustached man is my best bet in each group, he should still be the best bet. But I'd better check out the combined party to make sure.

Miss Lonelyhearts: So—when I attend the East Room party, I'll go after the men with mustaches!

When she finished her new chart, she was flabbergasted. The proportions had changed places! Now her best bet was a man *without* a mustache!

Miss Lonelyhearts: I had to change my tactics, and they worked! But I *still* don't understand it!

The curious paradox is easily modeled with playing cards. Red cards stand for swingers, black cards for squares. A large X on the back of a card symbolizes a mustache. No X on the back indicates no mustache.

Put an X on the back of five red cards and six black cards. To these cards add three red and four black that do not have X's. Thus there are 18 cards altogether. They represent the men in the East Room party.

Shuffle the 18 cards and spread them backs up. If you wish to maximize your chances of drawing a red card, should you pick one with an X or one without an X? It is easy to calculate the odds, as shown in the pictures, to see that your chance of picking a red card is best if you take an X card.

The men in the West Room party are modeled in the same way. Put X's on the backs of six red cards and three black cards. To these cards add nine red and five black without X's. There are 23 cards in all. Shuffle and spread backs up. Again, it is easy to show that if you wish to draw a red card, your success is maximized by taking an X card.

Now combine the two sets into a deck of 41 cards. Shuffle and spread. It is hard to believe, but if you perform the calculations correctly you will find that, if you wish to draw a red card, your chances now are higher if you select a card *without* an X!

Paradoxes like this can arise when statisticians analyze such data as the result of drug testing. For example, let the cards stand for the persons who participated in two research tests. Let X indicate a person given the drug, no X indicates a person given a placebo. Red cards are persons who improved, black cards are persons who did not. Each test, separately analyzed, will indicate that the drug had a more favorable effect than the placebo. But when the results of both tests are combined, the analysis indicates that the placebo had the more favorable effect! The paradox shows how difficult it is to design tests for which the statistical results are always trustworthy.

An instance of this paradox occurred at the University of California, Berkeley, in 1973, in connection with a study of possible sex bias in graduate school admissions. About 44 percent of men applying for graduate work were admitted, whereas only 35 percent of women were admitted. Since the qualifications of the men and women were roughly the same, this seemed a clear case of sex bias.

However, when the same data were examined to determine in which departments the discrimination occurred, it turned out, in essence, that in each department women had a greater chance of being accepted than men! How can this be explained? The paradox arose because a far higher percentage of women sought graduate work in difficult subjects that had high rejection rates. Taken major by major, a woman's chance to do graduate work was better than a man's. Only when all the data were combined did the bias swing the other way. Was the university exonerated by uncovering the source of the paradox? Perhaps, but one wonders if a scheme might be devised to make it more difficult to do graduate work in those subjects women tend to choose.

Hempel's Ravens

A famous paradox about black crows shows that Miss Lonelyhearts is in good company. Even the experts are still trying to understand it.

If only three or four crows are observed to be black, the scientific law that says: "All crows are black" is *weakly* confirmed. If millions of crows are seen to be black, it is *strongly* confirmed.

Crow: Caw, caw! I'm a *non-black* crow. As long as they never find me, they'll never know their law is false.

What about a yellow caterpillar? Could it be a confirming instance of the law?

To answer this, let's first state the law in a different but logically equivalent form: "All non-black objects are not crows."

Scientist: Aha! I've found a non-black object—a yellow caterpillar. It's definitely not a crow, therefore it confirms the law: "All non-black objects are not crows." So it must also confirm the equivalent law: "All crows are black."

It is easy to find millions of non-black objects that are not crows. Are they, too, confirming instances of the law: "All crows are black"?

Professor Carl Hempel, who invented this famous paradox, believes that a purple cow actually *does* slightly increase the probability that all crows are black. Other philosophers disagree. What's your opinion?

This is the most notorious of many paradoxes about confirmation theory that have been discovered in recent times. "The prospect of being able to investigate ornithological theories without going out in the rain," remarks Nelson Goodman (see the next paradox), "is so attractive that we know there must be a catch to it."

The problem is to find the catch. Hempel's belief is that observing a non-black object that is not a raven actually does confirm "all ravens are black," but only to an infinitesimal degree. Consider the testing of a hypothesis about a small number of objects, such as ten playing cards that are face down on a table. The hypothesis is that all the black cards are spades. We start turning the cards face up, one by one. Clearly, each time we turn over a black spade we have found a confirming instance.

Now we express the same hypothesis in different words: "All non-spade cards are red." Each card we turn that is not a spade, and that also is red, certainly confirms the theory as first stated. Indeed, if the first card is a black spade and all the other nine cards are red non-spades, we know the hypothesis is true.

The reason why this procedure seems strange when we apply it to non-ravens that are not black, says Hempel, is that the class of objects on earth that are not ravens is so enormously large compared to the number of ravens that the degree to which a non-raven that is not black confirms our hypothesis is negligible. Moreover, if we look around a room for non-ravens, already knowing there are no ravens in the room, we should not be surprised to find no non-black ravens in the room.

Yet if we do not have such additional knowledge, finding a non-raven that is non-black does, in a theoretical sense, count as a confirming instance of the hypothesis that ravens are black.

Opponents of Hempel like to point out that finding, say, a yellow caterpillar or a purple cow, by the same reasoning, must also be a confirming instance of the law, "all ravens are white." How can the same object confirm "all ravens are black" and "all ravens are white" simultaneously? The literature on Hempel's paradox is enormous; the paradox has a central role in the debate about confirmation of knowledge, which is the subject of the *Scientific American* article "Confirmation" by Wesley C. Salmon (May 1973).

Goodman's Grue

Another famous paradox of confirmation theory is based on the fact that many objects change color at some point in time. Green apples ripen to red, hair turns white in old age, silver tarnishes.

The Hempel and Goodman paradoxes show how little we understand the precise way in which statistics enters into the scientific method. We *do* know that without this invaluable tool, science could not continue its eternal quest for the laws that control our mysterious universe.

Nelson Goodman calls an object "grue" if it fulfills two conditions. First, it is green until the end of this century. Second, it is blue after that.

Now consider two different laws: "All emeralds are green," and "all emeralds are grue." Which law is best confirmed?

Strangely enough, both are equally confirmed! Every observation ever made of an emerald is a confirming instance of each law, and no one has ever observed a counterinstance! It is not easy to explain exactly why one law is accepted and the other is not.

Time

KIM 1981

Paradoxes about motion, supertasks, time travel, and reversed time

From the smallest subatomic particle to the largest galaxy, the universe is in a constant state of change, its incredible patterns altering every microsecond in the inexorable "flow" of time. (I put "flow" inside quotation marks because it is really the universe that flows. To say that time flows is as meaningless as saying length extends.)

It is hard to imagine an actual world without time. An object that existed only for zero seconds would not exist at all. Or could it? In any case, the flow of the universe is uniform enough to permit measurements, and with measurements come numbers and equations. Pure mathematics may be thought of as "timeless," but in applied mathematics, from simple algebra to calculus and beyond, vast areas deal with problems in which time is a fundamental variable.

This chapter brings together a variety of famous paradoxes about time and motion. Some of them, such as Zeno's paradoxes, were hotly debated by the ancient Greeks. Others such as the "dilation" of time in relativity theory, and the infinity machines that perform "supertasks," are products of this century. All of them should whet your appetite for paradoxes and for mathematics as well.

Here are some of the ways in which the paradoxes can serve as jumpoff points into serious mathematics and science:

The bicycle wheel paradox involves the cycloid curve, a splendid introduction to curves more complicated than the conic-section curves.

The frustrated skier dramatizes the power of simple algebra to prove an unexpected result.

Zeno's paradoxes, the rubber rope, supertasks, and the trotting dog all introduce the concept of limit, so essential to the understanding of calculus and all higher mathematics. Their resolution relies on Georg Cantor's theory of infinite sets, which we encountered in Chapter 2.

The worm on the rope is solved by using a famous series, the harmonic series.

The paradoxes about backward time, tachyons, and time travel introduce fundamental concepts essential to the understanding of relativity theory.

The trick for avoiding time-travel paradoxes, by assuming forking paths and parallel worlds, will introduce you to a strange approach in quantum mechanics called the "many-worlds interpretation."

The final paradox about the conflict between determinism and indeterminism offers a brief glimpse into one of the great perennial problems of philosophy.

Carroll's Crazy Clocks

Which clock keeps the best time? A clock that loses a minute a day or one that doesn't run at all?

Lewis Carroll argued this way:

Carroll: The clock that loses a minute a day is correct once every 2 years. The stopped clock is correct *twice* every 24 hours. So the stopped clock keeps the best time. Do you agree?

Alice is puzzled.

Alice: I know the stopped clock is right whenever it's 8 o'clock. But how do *I* know when it's exactly 8?

Carroll: That's easy to answer, my dear. You just stand by the stopped clock with a pistol in your hand.

Carroll: Keep your eyes on the clock. At the precise instant the clock is correct, fire the pistol. Everyone who hears the shot will know it is *exactly* 8.

Lewis Carroll was the pen name of Charles L. Dodgson, who taught mathematics at Christ Church, one of the colleges of Oxford University, England. His account of the two clocks can be found in *The Complete Works of Lewis Carroll* and in many other collections of Carroll's writings.

How did Carroll determine how often the slow-running clock is correct? Since the clock loses a minute a day, it will be correct again after losing 12 hours, which it does in 720 days.

The Perplexing Wheel

Lewis Carroll's clock paradox is just a nonsense joke, but here is one that isn't. Did you know that the tops of bicycle wheels move faster than the bottoms?

That's why the spokes in the upper half are blurred when a bicycle goes by.

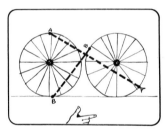

Let's look at two positions of the wheel as it moves along. Point A near the top has gone much farther than point B near the bottom. Speed is distance traveled per unit time, therefore point A has gone much faster than point B. Right?

When the speeds of the tops and bottoms of rolling wheels are compared, it is, of course, their ground speeds that one has in mind. One of the best ways to explain this paradox is to consider the curve called the cycloid. The cycloid curve is generated by any point on the rim of a wheel as the wheel rolls along a straight line. When the point is touching the ground, its speed is zero. As the wheel rolls, the point's speed accelerates, reaching its maximum when it is the top of the wheel. It then decelerates until it returns to zero speed when it touches the ground again. On wheels with flanges, such as the wheels of trains, a point on the flange actually moves *backward* in a tiny loop below the level of the track.

The cycloid has many beautiful mathematical and mechanical properties that are discussed in Chapter 13, "The Cycloid: Helen of Geometry" in my *Sixth Book of Mathematical Games from Scientific American*. The chapter explains how to draw a cycloid by rolling a coffee can. Constructing this curve and working out its equation can give better appreciation of its elegance and its unusual properties.

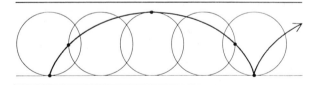

The Frustrated Skier

Skier: What a great day for skiing! I sure wish this lift moved faster than 5 kilometers per hour.

If the skier wants to raise his average speed to 10 kilometers per hour for the round trip up and down the slope, how fast must he ski down?

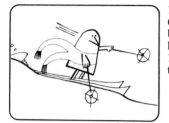

15 kilometers per hour? 60? 100? It's hard to believe, but the only way he can raise his average to 10 is by skiing down in *zero* time!

At first you may think that this paradox depends on the distance up and down the slope. This variable, however, is not relevant to the problem. The skier goes a certain distance up the slope at a certain speed. He wants to come down at a speed that will double his average speed for the round trip. But in order to do this, he must go *twice* the original distance in the *same* length of time it took him to go up. Clearly, to do this he must come down the slope in *no time at all*. Since this is impossible, there is no way that he can raise his average speed from 5 to 10 kilometers per hour. This is easily proved by elementary algebra.

Zeno's Paradoxes

The ancient Greeks invented many paradoxes about time and motion. One of the most famous is Zeno's argument about a runner.

Zeno devised a famous paradox about Achilles. The warrior wants to catch a turtle 1 kilometer away.

Zeno's runner reasoned:
Runner: Before I get to the finish line, I must pass the halfway point. Then I must reach the 3/4 mark, which is half the remaining distance.

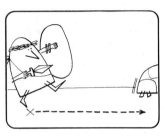

When Achilles gets to the spot where the turtle had been, the turtle has moved ahead about 10 meters.

Runner: And before I run the final quarter, I have to reach another halfway mark. These halfway marks never end. I'll *never* get there!

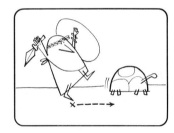

But when Achilles goes the 10 meters, the turtle has moved ahead again.
Turtle: You'll never catch me, old pal. Whenever you reach the spot where I last was, I'll always be ahead by *some* distance, even if it's less than a hair!

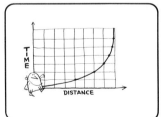

Suppose the runner takes 1 minute to go each half-segment. This time–distance graph shows how he gets closer and closer to the goal but never reaches it. Could his argument be right?

Zeno knew, of course, that Achilles *could* catch the turtle. He was simply showing the paradoxical consequence of viewing time and space as made up of an infinite number of discrete points that follow one another like beads on a string.

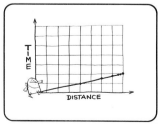

No, because the runner does *not* take a minute for each half-segment. Each segment is run in half the time it takes for the preceding one. He'll reach the goal in just 2 minutes even though he must pass an infinity of halfway points.

In both of these paradoxes we must think of the runners as being equivalent to points moving with uniform speed along a straight line. Zeno knew that a point moving from A to B does indeed get there. Zeno's paradoxes were designed to bring out the difficulty encountered when one tries to explain motion by breaking a line into distinct points that lie "next" to one another, and time into distinct instants that follow "after" one another.

Merely to show, as we did, that the runner does reach B because the time required to run each new half-segment is half the time required to run the previous one, would not have satisfied Zeno. He would have replied that just as there is always another halfway point on the line to be reached, so there is always another halfway instant of time to be reached. In brief, the argument Zeno applied to the line can also be applied to the time sequence. The time gets closer and closer to 2 minutes, but there always remains an infinity of instants yet to go. The same is true of the paradox of Achilles and the turtle. At every step in the infinite process, there always remains an infinity of "next" steps to be made in both space and time.

Many philosophers of science agree with Bertrand Russell's famous discussion of Zeno's paradoxes in the sixth lecture of his book *Our Knowledge of the External World.* Russell argues that Zeno's paradoxes were not *effectively* answered until Georg Cantor developed his theory of infinite sets.

Cantor's theory allows one to treat infinite sets of points in space, or events in time, as completed wholes rather than simply a collection of isolated individual points and events. At the heart of Zeno's paradoxes is the impossibility of viewing segments of space and time as made up of an infinity of members that nevertheless are as discrete and separate from one another as footsteps in the snow. The resolution of his paradoxes demands a theory like Cantor's, which unites our intuitive notion of individual points and events with a systematic theory of infinite sets.

The Rubber Rope

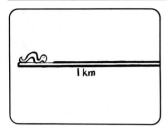

Here's a new paradox that Zeno didn't think of. A worm is at one end of a rubber rope. The rope is 1 kilometer long.

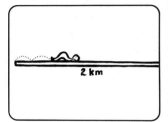

The worm crawls along the rope at a steady pace of 1 centimeter per second. After the first second, the rope stretches like a rubber band to 2 kilometers. After the next second, it stretches to 3 kilometers, and so on. Will the worm ever reach the end of its rope?

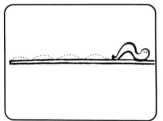

Your intuition tells you that the worm will *never* reach the end. But it does! How long does it take?

The key to this problem is understanding that the rope stretches uniformly like a rubber band. This means that the worm is *carried forward with the stretching*.

A good way to solve the puzzle is to measure the worm's progress after each second as a fraction of the rope's length after that second. When the sum of these fractions is 1, the worm has come to the end of its rope.

There are 100,000 centimeters in a kilometer, so at the end of the first second the worm has traveled (1/100,000)th of the rope's length. After the next second the worm crawls forward another centimeter. This distance covers an additional (1/200,000)th of the rope's new length of 2 kilometers. After the third second the worm has gone an additional (1/300,000)th of the rope's length of 3 kilometers, and so on. After k seconds the worm's progress, expressed as a fraction of the entire rope, is

$$\frac{1}{100,000}\left(\frac{1}{1} + \frac{1}{2} + \frac{1}{3} + \frac{1}{4} + \cdots + \frac{1}{k}\right)$$

The series inside the parentheses is known as the harmonic series. Notice that the sum of the terms from 1/2 through 1/4—that is, the sum of 1/3 and 1/4—exceeds $2 \times 1/4 = 1/2$. Similarly, the sum of the terms from 1/4 out through 1/8 exceeds $4 \times 1/8 = 1/2$. Thus the sum of the series from 1/1 out through $1/2^k$ always exceeds $k \times 1/2 = k/2$, as you can see by bunching the terms. First take the sum of two terms, then the sum of the next four, then the next eight, and so on. The partial sum of the harmonic series can be made as large as one desires.

The worm will reach the end of the rope before $2^{200,000}$ seconds. A more refined estimate is $e^{100,000}$ seconds, where e is the basis of the natural logarithms (e is an irrational number slightly larger than 2.7). This gives both the elapsed time in seconds and the rope's length in kilometers.

For the precise formula for determining a partial sum of the harmonic series, see "Partial Sums of the Harmonic Series" by R. P. Boas, Jr., and J. M. Wrench, Jr., in *American Mathematical Monthly* (vol. 78, October 1971, pp. 864–870). The final length of the rope proves to be enormously longer than the diameter of the known universe, and the time it takes the worm to reach the end vastly exceeds the estimated age of the universe. Of course, the problem is about an idealized worm that represents a point on an idealized rope. A real worm would die after barely getting started on the trip, and a real rope would have to stretch so thin that it would consist of molecules separated by inconceivably vast spaces.

Regardless of the problem's parameters, which are the initial length of the rope, the worm's speed, and how much the rope stretches after each unit of time, the worm always reaches the rope's end in a finite time. Good problems arise by altering the way the rope stretches. For example, what happens if the rope stretches in a geometric progression, say by doubling its length after each second? In this case, the worm never gets to the rope's end.

Supertasks

Philosophers are now arguing about a new class of time paradoxes called supertasks. One of the simplest involves a lamp. A pushbutton turns it on and off.

The lamp is turned on for 1 minute, then off for 1/2 minute, on for 1/4 minute, and so on. This series ends in just 2 minutes. After it ends, will the lamp be on or off?

Every odd push of the button turns the lamp on. Every even push turns it off. If the lamp is on at the finish, it means the last counting number is odd. If off at the finish, the last number is even. But there *is* no last counting number. The lamp *must* be on or off, but there is no way to know which!

Philosophers of science are not yet agreed on how to clear up paradoxes involving "supertasks"— tasks performed by what are called "infinity machines." The lamp paradox is known as the Thomson lamp after James F. Thomson who first wrote about it. Everyone agrees that a Thomson lamp cannot be constructed, but that's not the point. The point is whether the lamp is logically conceivable if certain assumptions are made. Some argue that the lamp is a meaningful "thought experiment." Others contend that it is nonsense.

The paradox is disturbing because there seems to be no logical reason why the lamp, like Zeno's runner, cannot complete an infinite sequence of ons and offs. If Zeno's runner can cover an infinity of halfway points in 2 minutes, why cannot the lamp's idealized switch be turned an infinity of times to end the sequence in *exactly* 2 minutes? But if the lamp *can* do this, it seems to prove that there is a "last" counting number, which is absurd.

The philosopher Max Black has given the same paradox in the form of an infinity machine that transfers a marble from tray A to tray B in 1 minute, then in 1/2 minute it puts the marble back on tray A, and in the next 1/4 minute it puts the marble on B, and so on in the same time-halving series as before. This series converges and ends precisely after 2 minutes. Where is the marble? If it is on either tray, it implies that the last counting number is odd or even. Since there is no last counting number, both possibilities seem to be eliminated. But if the marble is not on a tray, where is it?

If you are interested in supertasks, you will find the basic papers reprinted in *Zeno's Paradoxes,* a collection of essays edited by Wesley C. Salmon, and the paradoxes analyzed at length in Adolf Grünbaum's *Modern Science and Zeno's Paradoxes.*

Mary, Tom, and Fido

Here's a supertask performed by a dog. At the start, Fido is with his master Tom. Mary is 1 kilometer away.

Mary: Fido trots at 8 kilometers per hour, so in a quarter of an hour he's gone a quarter of that distance, or 2 kilometers.
Tom: By golly, you're right! I don't even need this calculator!

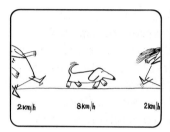

Tom and Mary walk toward each other at 2 kilometers per hour. Fido, who loves them equally, trots back and forth between them at 8 kilometers per hour. Assume that he makes each turn instantaneously.

Suppose that Tom, Mary, and Fido start at the middle of the same path. Tom and Mary walk backward at the same rate as before, while Fido trots back and forth between them. When Tom and Mary reach the ends of the path, where is Fido?

Fido's path is easy to follow on this time–distance graph. When Tom and Mary meet at the center, is Fido facing Tom or Mary?

It seems impossible, but the dog can be *anywhere* between Tom and Mary! If you don't believe it, put Fido at *any* spot between them and start the event forward in time. At the finish, all three will be together at the center.

That question is as impossible to answer as whether the lamp is on or off. But *we can* help Tom figure out how far the dog has run.
Tom: Darn it, Mary, I've got to sum a complicated series of zigs and zags.

Mary: No you don't, you dummy. We each walk 2 kilometers per hour, so we each go half a kilometer in 15 minutes. Since we started 1 kilometer apart, we met at the end of 15 minutes.

The first problem, in which Mary and Tom walk toward each other while Fido trots back and forth between them, is a classic problem that has many different story lines. Sometimes it is a bird that flies back and forth between two approaching locomotives, sometimes a fly that buzzes back and forth between two approaching bicycles.

A story is told about the eminent Hungarian mathematician, John Von Neumann. Someone gave him a version of this problem. Von Neumann thought a moment before he supplied the correct answer. The person who presented the problem congratulated him. "Most people," he said, "think they have to solve it the hard way by summing an infinite series of path segments." Von Neumann looked surprised. "But that's what I did," he said.

Which way is Fido facing by the time Mary and Tom meet? This is similar to asking whether the Thomson lamp is on or off, or whether the marble is in tray A or B. It seems as if the dog must be facing either Tom or Mary, but in each case the answer implies that the last counting number, applied to the infinite sequence of zigs and zags, is either odd or even.

When we time-reverse the process by starting with Mary, Tom, and Fido in the center of the path, and move Mary and Tom backward while Fido runs back and forth as before, another paradox arises. Our intuition tells us that if a well-defined procedure is time-reversed, in the sense that all the motions go the other way, we must end exactly as we started. The curious thing about this case is that the procedure is no longer well-defined when it is time-reversed. When the event goes forward in time, it ends with Fido exactly at the center. But when the same event is run backward, Fido's position at the finish cannot be determined. The dog can be at *any* point along the path.

For a more detailed discussion of this paradox, see Wesley Salmon's analysis in the Mathematical Games department of *Scientific American,* December 1971. This problem and the previous paradoxes about supertasks and runners are descriptive introductions to the concept of limit as well as applications of summing a geometric series.

Fido's zigzag path is similar to the path of a bouncing ball. Here is a simple bouncing ball problem. Suppose an ideal ball is dropped from a height of 1 meter. It always bounces to one-half its previous height. If each bounce takes a second, the ball will bounce forever. But, like Zeno's runner, the lamp, the marble machine, and Fido, each segment of the ball's path is covered in less time than the previous one. In this case each successive bounce takes $1/\sqrt{2}$ times the length of the previous bounce. The time sequences also converge to a limit, which means that the ball stops bouncing after a finite time, even though it makes (in theory) an infinite number of bounces. The ball travels a distance of $1 + 1/2 + 1/4 + \cdots + 1/n = 2$ meters.

Suppose the ball always bounces to one-third its previous height. How far does it travel before it comes to rest?

Can Time Go Backward?

When certain motions are reversed, such as a person walking backward or a car going backward, it almost seems as if time has been reversed.

We cannot see the future, but we *can* look into the past. When you see a star a thousand light-years away, you are seeing it the way it looked a thousand years ago.

This familiar song . . .

But seeing the past is not the same as entering it. Will it ever be possible to get into a time machine and actually *visit* the past or future?

. . . sounds funny when it is played backward.

Most events in life are impossible to reverse.

Time is like an arrow that always points in the same direction. Even when a song is played backward, the notes still follow one another in forward time.

Consider what kinds of events can be "time-reversed" in the sense of reversing the direction of motion, and what kinds cannot. A good way to make the distinction clear is to suppose that an event is being photographed by a motion picture camera. Later, the picture is shown on a screen, but the film is run backward. What sort of events will seem to violate natural law when run backward? What sort will not?

For example, a motion picture of a car moving backward does not appear impossible. Perhaps the driver is simply backing up his car. But a motion picture of a diver coming feet first out of the water and going back up to a diving board is immediately recognized as a sign that a film has been reversed. The same is true of a motion picture that shows a broken egg coming together again on the floor and hopping up to a person's hands. Such an event could never happen in the actual world.

Even when an event is "time-reversed" by changing the direction of motion, like playing a record backward, the event is still going forward in time. Arrows normally move in the direction they are pointing. Suppose you saw an arrow travel backward through the sky and end in an archer's bow. It would reach the bow at a later time than when it was in midair. Sir Arthur Eddington once compared time to a symbolic arrow that always points in the same direction. Events in our universe seem relentlessly to go from past to future, never from future to past.

In recent years physicists and cosmologists have been speculating about the possibility of events going the "other way" in other universes. And there is an interpretation of quantum mechanics, by Nobel Laureate Richard Feynman, in which antiparticles are regarded as particles momentarily going backward in time! You can read about these fantastic speculations in the last four chapters of the second edition of my *Ambidextrous Universe*.

Time Machines

Professor Brown has just gone back 30 years in time. He is looking at himself when he was a baby.
Brown: Suppose I killed this baby. Then there would be no one to grow up and become Professor Brown! Would I suddenly vanish?

Now Professor Brown has traveled 30 years into the future. He is carving his name on an oak tree outside his laboratory.

Brown was here

The professor returned to the present, and a few years later decided to chop down the oak tree. When he finished, he became very perplexed.

Brown: Hmmm. Three years ago I went 30 years into the future and carved my name on this tree. What will happen, 27 years from now, when I arrive from the past? There won't *be* any tree. Where did that tree come from that I carved my name on?

Hundreds of science fiction stories, motion pictures, and television shows have been written about time travel to the past or future. The classic story of this type is *The Time Machine* by H. G. Wells.

Is time travel *logically* possible or does the notion lead to contradictions? It is clear from the paradoxes that if we assume there is one single universe, moving forward in time, any attempt to enter the past can lead to a logical absurdity. Consider the first paradox, in which a time traveler enters his own past and sees himself as a baby. If he kills the baby, he will both exist and not exist. If the baby, which grew up to become Professor Brown, is killed, then where does Professor Brown come from?

The second paradox is more subtle. There is no contradiction about Professor Brown going forward in time and carving his name on the tree. The contradiction arises after he has returned to the present—that is, after he has gone *backward* in time. By chopping down the tree, he eliminates it from the future. So we have a contradiction again. At a certain time in the future, the tree both exists and does not exist.

The Tachyon Telephone

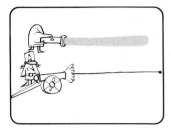

In recent years physicists have speculated about sub-atomic particles called *tachyons*. Tachyons move faster than light. According to relativity theory, if tachyons exist they must also move backward in time.

Professor Brown thinks he has invented a tachyon telephone for communicating with his friend, Dr. Gamma, in another galaxy.

Dr. Brown is telling his students about an experiment:
Brown: Tomorrow at noon I will ring Dr. Gamma on my tachyon phone. I'll ask him to hang up, count the number of helicopters outside his window, then call me back with the number.
Assistant: It won't work, sir.

Brown: And why not, young lady?
Assistant: Because tachyons go back in time. Dr. Gamma will get your call an hour *before* noon. His return call will go back another hour, so you'll get your answer 2 hours *before* you ask the question! That's not possible.

This episode proves it is not necessary for a *person* to move back in time to generate a paradox. If any sort of message or object is sent back in time, contradictions can arise. For example, Professor Brown might say to himself on Monday: "Next Friday I will put my necktie in this time machine and send it back to Tuesday, which is tomorrow." Sure enough, on Tuesday he finds his tie in the machine. Suppose he then burns the tie. When Friday arrives, there will be no tie to send back. Once more, the tie seems to both exist and not exist on Friday. It existed when Professor Brown sent it back to Tuesday, but now it is Friday again and there is no tie to send back!

Tachyons, however, are taken quite seriously by many physicists. (See "Particles That Go Faster Than Light" by Gerald Feinberg, *Scientific American,* February 1970.) According to relativity theory, the speed of light is an upper limit for ordinary particles. Physicists have speculated, however, on the possible existence of particles, which Feinberg named tachyons, that always move much faster than light. For tachyons, the speed of light is a *lower* limit. Relativity theory makes it necessary to assume that such particles, if they exist, must move backward in time like the lady in the familiar limerick:

> There was a young lady named Bright
> Who traveled much faster than light.
>   She set out one day
>   In her relative way,
> And returned on the previous night.

The telephone paradox does not prove that tachyons cannot exist, but it does show that *if* they do, there is no way they could be used for communication. If there were, we would have the logical contradiction explained above. For more on this paradox and its implications for tachyon research, see G. A. Benford, D. L. Book, and W. A. Newcomb, "The Tachyon Antitelephone," (*Physical Review*, D, vol. 2, July 15, 1970).

Parallel Worlds

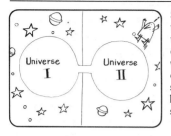 Science fiction writers have thought of a fantastic way to avoid time travel paradoxes. They imagine that whenever a time traveler enters the past, the universe splits into two identical halves, each in a different space-time.

 This forking universe theory has a lot of strange possibilities. Suppose you go back one year and shake hands with yourself:

Feemster: Hi, Feemster.
Feemster: Glad to meet you, Feemster.

 Here's how it works. Suppose you go back to 1930 and shoot Hitler. As soon as this happens, the universe divides into parallel worlds or timelines.

 Any time later, either of you could hop in the time machine again and go back to meet *two* duplicates of yourself. Now there are *three* Feemsters. By repeating this, hundreds of Feemsters could be created.

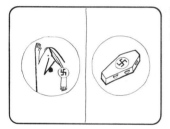 Universe I goes on with Hitler alive. Universe II goes on with Hitler dead.

 If you return to the present of Universe II you'll find old newspapers telling how Hitler was killed. The world you left, in which Hitler was *not* killed, is a world to which you can never return.

The pictures describe one fantastic method of permitting backward time travel without encountering logical contradictions. Science fiction writers were the first to think of it, and scores of science fiction stories have been based on it. The trick is to assume that whenever a person or thing enters the past, the universe splits into parallel worlds. If this occurs, there is no longer a contradiction between Professor Brown both existing and not existing, or between the tree existing and not existing. If there are parallel worlds, Brown (or the tree) may exist in one but not the other.

Amazingly, there is an interpretation of quantum mechanics based on this concept of forking universes. Called the "many-worlds theory," there is an entire book about it: *The Many-Worlds Interpretation of Quantum Mechanics,* edited by Bryce S. DeWitt and Neill Graham. According to this wild theory, first advanced in 1957 by Hugh Everett III, the universe branches at *every* microsecond into countless parallel worlds, each a possible combination of microevents that *could* occur at that instant. This leads to an incredible vision of an infinity of universes that represent every possible combination of possible events. As Frederic Brown described the vision in his science fiction novel, *What Mad Universe:*

> If there are infinite universes, then all possible combinations must exist. Then, somewhere, *everything must be true.* . . . There is a universe in which Huckleberry Finn is a real person, doing the exact things Mark Twain described him as doing. There are, in fact, an infinite number of universes in which a Huckleberry Finn is doing every possible variation of what Mark Twain *might* have described him as doing. . . . And infinite universes in which the states of existence are such that we would have no words or thoughts to describe them or to imagine them.

Time Dilation

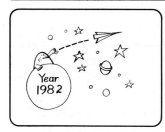

Traveling into the past creates such wild paradoxes that no scientist takes it seriously. But travel into the future is another matter. Suppose a spaceship leaves earth and travels at almost the speed of light.

This kind of time travel does not lead to paradox, but the astronauts are now trapped in earth's future. They cannot come back.

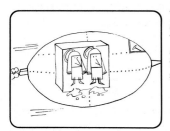

The faster a spaceship travels, the slower its time goes. Time would seem normal to astronauts in the ship, but to us they would seem like statues.

The spaceship goes to another galaxy and returns. For astronauts on the ship, the trip seems to last only five years. But when the ship lands back on earth, thousands of earth years will have gone by!

Contradictions arise only from travel into the past, not into the future. After all, we are all time travelers moving into the future whether we like it or not. When you go to sleep at night you expect to wake up in the near future. A person could be placed in suspended animation and be revived, say, a thousand years later. Many science fiction novels and stories have been based on this kind of "time travel," notably *When the Sleeper Wakes* by H. G. Wells.

As our cartoon panels show, a quite different way of traveling into the future is provided by Einstein's theory of relativity. According to the special theory of relativity, the faster an object moves, the slower its time goes relative to a stationary observer. For example, if a spaceship has a speed close to that of light, time on the ship is much slower than time on earth. On the ship, astronauts would not be aware of anything unusual. Their clocks would seem to run normally, their hearts would beat at the usual rate, and so on. But if there were any way that people on earth could observe them, they would seem to be moving so slowly that they would appear like statues. If the astronauts in turn could observe life on earth, events would seem to be moving so rapidly that an earth year would go by in just a few hours.

The reason we do not observe these effects in everyday life is that they become significant only at velocities close to the speed of light, conventionally symbolized by c, which is about 186,000 miles per second. The simple formula relating the length of time, T, measured by earthbound clocks, compared to the interval T', measured by clocks on a spaceship traveling at a constant velocity v with respect to the earth is

$$T' = \frac{T}{\sqrt{1 - \dfrac{v^2}{c^2}}}$$

Substitute any commonly encountered velocity for v in the expression under the radical sign and you will get a value for this expression so close to 1 that T and T' will be essentially equal. But if you give v a value of $.5c$ or $.75c$ or $.9c$ (velocities that are reached by high-speed subatomic particles), the time dilation becomes large enough to measure in the laboratory. Such measurements provide strong confirmations of the special theory of relativity.

Fate, Chance, and Free Will

Although physicists are learning more and more about time, its essence remains a dark mystery. One of the biggest questions is whether the future is completely determined.

Determinist: *Que sera, sera.* Whatever will be, will be. Life is like a movie. We are the creatures on the screen. We *think* we have free wills. Actually, we are just acting out predetermined events.

Indeterminist: The future is only *partly* determined. We can change things by using our will. History has genuine surprises.

Scientists, philosophers, and ordinary people divide sharply over the question of whether the future is completely determined by the past. A determinist believes that the total state of the universe at any given moment fully determines the total state of the universe at any future moment. This was Einstein's personal belief. One of the greatest of all philosophers who espoused determinism was Benedict de Spinoza, and Einstein considered himself a Spinozist. This was one reason why Einstein could never accept quantum theory as being final, because in quantum theory chance plays a fundamental role in determining events on the microlevel. "I do not believe God plays dice with the universe," was how Einstein once expressed it.

An indeterminist believes that the future of the universe is determined only in part by its present state. He need not believe in free will. He may believe no more than that the role of chance on the microlevel prevents the future from being completely determined. In addition, he may also believe that living creatures, and especially humans, possess "free will" that gives them the power to alter significantly the future in ways that could not be predicted even by a superbeing who knew everything there was to know about the universe's present state. Charles Peirce and William James were two eminent American philosophers who championed the indeterminist cause.

These profound philosophical questions are intimately bound up with the nature of time and with what is meant when we say that one event "causes" another. No one doubts that mathematics can be applied to our measurements of the universe in such a way that many events can be predicted with almost perfect accuracy: the time of the next solar eclipse, for example. And no one denies that other events, such as the next fall of a die or what the weather will be next week, cannot in practice be predicted precisely because the causal factors are too complex.

The big question is whether the basic laws of the universe are completely deterministic, or whether genuine novelty is created by pure chance on the microlevel, or by the free will of living creatures on the macrolevel, or perhaps by both. These questions were debated by the ancient Greeks, and scientists, philosophers, and everyone else have been debating them ever since.

References and Suggested Readings

Suggested readings are preceded by ★.

1 Logic

General References

Carroll, Lewis. *The Annotated Alice: Alice's Adventures in Wonderland and Through the Looking Glass.* Martin Gardner, ed. New York: Clarkson N. Potter, Bramhall House, 1960.

★ Dunsany, Lord. *The Ghost of the Heaviside Layer and Other Fantasies.* Philadelphia: Owlslick Press, 1980.

Fisher, John. *The Magic of Lewis Carroll.* New York: Simon and Schuster, 1973.

★ Hofstadter, Douglas R. *Gödel, Escher, Bach: An Eternal Golden Braid.* New York: Basic Books, 1979.

★ Quine, W. V. "Paradox." *Scientific American*, April 1962.

Russell, Bertrand. *Principia Mathematica*, Part 8. Cambridge: Cambridge University Press, 1910–1913.

Russell, Bertrand. *My Philosophical Development.* Reprint: Winchester, Mass.: Allen Unwin, 1975.

★ Smullyan, Raymond. *What Is the Name of This Book?* Englewood Cliffs, N.J.: Prentice-Hall, 1978.

★ Smullyan, Raymond. *This Book Needs No Title.* Englewood Cliffs, N.J.: Prentice-Hall, 1980.

★ van Heijenoort, Jean. "Logical Paradoxes." In *The Encyclopedia of Philosophy.* Paul Edwards, ed. New York: Macmillan, 1967.

The Liar Paradox

Martin, Robert L., ed. *The Paradox of the Liar.* New Haven: Yale University Press, 1970.

Tarski, Alfred. "Truth and Proof." *Scientific American*, June 1969.

Infinite Regress

Gardner, Martin. "Infinite Regress." Chapter 22 in *Sixth Book of Mathematical Games from Scientific American.* San Francisco: W. H. Freeman and Company, 1971.

Prediction Paradoxes

★ Gardner, Martin. "Mr. Apollinax Visits New York." Chapter 11 in *New Mathematical Diversions from Scientific American.* New York: Simon & Schuster, 1966.

★ Gardner, Martin. "The Paradox of the Unexpected Hanging." Chapter 1 in *The Unexpected Hanging and Other Mathematical Diversions.* New York: Simon & Schuster, 1968.

Newcomb's Paradox

Brams, Steven. "A Problem of Prediction." Chapter 8 in *Paradoxes in Politics: An Introduction to the Nonobvious in Political Science.* New York: Free Press, 1976.

★ Gardner, Martin. "Free Will Revisited." Mathematical Games Department, *Scientific American*, July 1973.

★ Nozick, Robert. "Newcomb's Problem and Two Principles of Choice." In *Essays in Honor of Carl G. Hempel.* Nicholas Rescher, ed. Atlantic Highlands, N.J.: Humanities Press, 1970.

★ Nozick, Robert. "Reflections on Newcomb's Problem." Mathematical Games Department, *Scientific American*, March 1974.

2   Number

General References

Beiler, Albert H. *Recreations in the Theory of Numbers*. New York: Dover, 1964.

Dantzig, Tobias. *Number: The Language of Science*, 4th ed. New York: Free Press, 1967.

Gardner, Martin, ed. *Scientific American Book of Mathematical Puzzles and Diversions*. New York: Simon & Schuster, 1963.

Northrop, Eugene. *Riddles in Mathematics: A Book of Paradoxes*. Huntington, N.Y.: Krieger, 1975.

Magic Tricks

Barr, George. *Entertaining with Number Tricks*. New York: McGraw-Hill, 1971.

★ Fulves, Karl. *Self-Working Card Tricks*. New York, Dover, 1976.

★ Gardner, Martin. *Mathematics, Magic and Mystery*. New York, Dover, 1956.

Transfinite Numbers

Cohen, Paul, and Reuben Hersh. "Non-Cantorian Set Theory." *Scientific American*, December 1967.

★ Gardner, Martin. "The Orders of Infinity." Mathematical Games Department, *Scientific American*, March 1971.

★ Gardner, Martin. "Aleph-Null and Aleph-One." Chapter 3 in *Mathematical Carnival*. New York: Knopf, 1975.

★ Kasner, Edward, and James Newman. "Beyond the Googol." Chapter 2 in *Mathematics and the Imagination*. New York: Simon & Schuster, 1940.

3   Geometry

General References

★ Anno, Mitsumasa. *Anno's Alphabet: An Adventure in Imagination*. New York: T. Y. Crowell, 1975.

★ Anno, Mitsumasa. *The Unique World of Mitsumasa Anno*. New York: Philomel Books, 1980.

★ Abbott, Edwin A. *Flatland: A Romance of Many Dimensions*. 1884. Various reprints available.

Burger, Dionys. *Sphereland*. Cornelie J. Rheinboldt, trans. New York: T. Y. Crowell, 1965.

Courant, Richard, and Herbert Robbins. *What Is Mathematics?* Oxford: Oxford University Press, 1941.

Coxeter, H. S. M. *Introduction to Geometry*. New York: Wiley, 1961.

★ Gardner, Martin. *Mathematics, Magic and Mystery*. New York: Dover, 1956.

Gardner, Martin. *Sixth Book of Mathematical Games from Scientific American*. San Francisco: W. H. Freeman and Company, 1971.

Gardner, Martin. *New Mathematical Diversions from Scientific American*. New York: Simon & Schuster, 1971.

Gardner, Martin. *Mathematical Circus*. New York: Vintage Books, 1981.

Gardner, Martin, ed. *Second Scientific American Book for Mathematical Puzzles and Diversions*. New York: Simon & Schuster, 1965.

Jacobs, Harold. *Geometry*. San Francisco: W. H. Freeman and Company, 1974.

Mandelbrot, Benoit. *The Fractal Geometry of Nature*. San Francisco: W. H. Freeman and Company, 1982.

Ogilvy, C. Stanley. *Excursions in Geometry*. Oxford: Oxford University Press, 1969.

Wells, H. G. *28 Science Fiction Stories*. New York: Dover, 1952.

Mirror Symmetry

★ Kim, Scott. *Inversions*. New York: McGraw-Hill, Byte Books, 1981.

Lockwood, S. H., and R. H. Macmillan. *Geometric Symmetry*. Cambridge: Cambridge University Press, 1978.

Weyl, Hermann. *Symmetry*. Princeton: Princeton University Press, 1952.

Topology

Arnold, Bradford. *Intuitive Concepts in Elementary Topology*. Englewood Cliffs, N.J.: Prentice-Hall, 1962.

Barr, Stephen. *Experiments in Topology*. New York: T. Y. Crowell, 1972.

★ Tucker, Albert, and Herbert Bailey, Jr. "Topology." *Scientific American*, January 1950.

Antimatter
Alfvén, Hannes. *Worlds-Antiworlds: Antimatter in Cosmology*. San Francisco: W. H. Freeman and Company, 1966.

★ Gardner, Martin. *The Ambidextrous Universe: Mirror Asymmetry and Time-Reversed Worlds*, 2nd ed. New York: Scribner's, 1979.

Yang, Chen Ning. *Elementary Particles: A Short History of Some Discoveries in Atomic Physics*. Princeton: Princeton University Press, 1962.

4 Probability

General References
Jacobs, Harold. *Mathematics: A Human Endeavor*, 2nd ed. San Francisco: W. H. Freeman and Company, 1982.

Kraitchik, Maurice. *Mathematical Recreations*, 2nd ed. New York: Dover, 1953.

Mosteller, Frederick. *Fifty Challenging Problems in Probability*. Reading, Mass.: Addison-Wesley, 1965.

Thorp, Edward. *Elementary Probability*. New York: Wiley, 1966.

★ Weaver, Warren. *Lady Luck: The Theory of Probability*. New York: Doubleday, 1963.

Gambling
★ Epstein, Richard. *The Theory of Gambling and Statistical Logic*. New York: Academic Press, 1967.

Jacoby, Oswald. *How to Figure the Odds*. New York: Doubleday, 1947.

Scarne, John. *Scarne's Complete Guide to Gambling*. New York: Simon & Schuster, 1961.

Pascal's Wager
Cargile, James. "Pascal's Wager." *Philosophy*, vol. 41, July 1966.

James, William. Chapters 1 and 3 in *The Will to Believe and Other Essays in Popular Philosophy*. London: Longmans Green, 1903.

Turner, Merle. "Deciding for God—The Bayesian Support of Pascal's Wager." *Philosophy and Phenomenological Research*, vol. 29, September 1968.

5 Statistics

General References
★ Huff, Darrel. *How to Lie with Statistics*. New York: Norton, 1954.

Levinson, Horace. *Chance, Luck and Statistics*. New York: Dover, 1963.

Moroney, M. J. *Facts from Figures*. New York: Penguin, 1956.

Mosteller, F. R., Robert E. Rourke, and G. B. Thomas, Jr. *Probability and Statistics*. Reading, Mass.: Addison-Wesley, 1961.

The Small-World Paradox
★ Gardner, Martin. "Why the Long Arm of Coincidence Is Usually Not as Long as It Seems." Mathematical Games Department, *Scientific American*, October 1972.

★ Milgram, Stanley. "The Small World Problem." *Psychology Today*, May 1967.

Birthdate Paradox
Goldberg, Samuel. "A Direct Attack on a Birthday Problem." *Mathematics Magazine*, vol. 49, May 1976, pp. 130–132.

Mosteller, Frederick. "Understanding the Birthday Problem." *The Mathematics Teacher*, May 1962, pp. 322–325.

Nontransitive Paradoxes
Black, Duncan. *The Theory of Committees and Elections*. Cambridge: Cambridge University Press, 1958.

★ Gardner, Martin. "On the Paradoxical Situations That Arise from Nontransitive Relations." Mathematical Games Department, *Scientific American*, October 1974.

Hempel's Ravens
★ Salmon, Wesley. "Confirmation." *Scientific American*, May 1973.

Schlesinger, G. "Hempel's Paradox." Chapter 1 in *Confirmation and Confirmability*. Oxford: Oxford University Press, 1974.

Goodman's Grue
Goodman, Nelson. *Fact, Fiction, and Forecast*. New York: Bobbs-Merrill, 1965.

Hesse, Mary. "Ramifications of 'Grue'." *British Journal for the Philosophy of Science*, vol. 20, May 1959, pp. 13–25.

6 Time

General References
Brown, Frederic. *What Mad Universe*. Mattituck, N.Y.: Amereon Ltd., 1976 (reprint of 1949 ed.).

DeWitt, Bryce S., and Neill Graham, eds. *The Many-Worlds Interpretation of Quantum Mechanics*. Princeton: Princeton University Press, 1973.

Gale, Richard, ed. *The Philosophy of Time: A Collection of Essays:* Atlantic Highlands, N.J.: Humanities Press, 1978.

Gardner, Martin. *Sixth Book of Mathematical Games from Scientific American*. San Francisco: W. H. Freeman and Company, 1971.

Gold, Thomas, ed. *The Nature of Time*. Ithaca, N.Y.: Cornell University Press, 1967.

Priestley, J. B. *Man and Time*. New York: Doubleday, 1964.

★ Whitrow, G. J. *The Natural Philosophy of Time*. New York: Harper & Row, 1961.

Zeno's Paradoxes and Supertasks
Grünbaum, Adolf. *Modern Science and Zeno's Paradoxes*. New York: Wesleyan University Press, 1967.

★ Salmon, Wesley C., ed. *Zeno's Paradoxes*. New York: Bobbs-Merrill, 1969.

The Direction of Time
★ Davies, P. C. W. *The Physics of Time Asymmetry*. Berkeley: University of California Press, 1974.

★ Gardner, Martin. *The Ambidextrous Universe: Mirror Asymmetry and Time-Reversed Worlds*, 2nd ed. New York: Scribner's, 1979.

Time Travel
Edwards, Malcolm. "Time Travel." In *The Science Fiction Encyclopedia*. Peter Nicholls, ed. New York: Doubleday, 1979.

★ Gardner, Martin. "On the Contradictions of Time Travel." Mathematical Games Department, *Scientific American*, May 1974.

Van Doren Stern, Philip, ed. *Travelers in Time*. New York: Doubleday, 1947.